Hyperparameter Tuning for Machine and Deep Learning with R

Eva Bartz · Thomas Bartz-Beielstein ·
Martin Zaefferer · Olaf Mersmann
Editors

Hyperparameter Tuning for Machine and Deep Learning with R

A Practical Guide

 Springer

Editors
Eva Bartz
Bartz & Bartz GmbH
Gummersbach, Germany

Martin Zaefferer
Bartz & Bartz GmbH and with Institute for
Data Science, Engineering, and Analytics,
TH Köln
Gummersbach, Germany

Duale Hochschule Baden-Württemberg
Ravensburg
Ravensburg, Germany

Thomas Bartz-Beielstein
Institute for Data Science, Engineering,
and Analytics
TH Köln
Gummersbach, Germany

Olaf Mersmann
Institute for Data Science, Engineering,
and Analytics
TH Köln
Gummersbach, Germany

ISBN 978-981-19-5172-5   ISBN 978-981-19-5170-1   (eBook)
https://doi.org/10.1007/978-981-19-5170-1

This Springer imprint is published by the registered company Springer Nature Singapore Pte Ltd.
The registered company address is: 152 Beach Road, #21-01/04 Gateway East, Singapore 189721,
Singapore

Foreword

Hyperparameter tuning? Is this relevant in practice? Is it not rather an academic gimmick? That the latter is not the case has been known for many years. On the other hand, it is mostly unclear what exactly this looks like in practice. Which procedures depend on which hyperparameters? How sensitive are the procedures to different settings of their hyperparameters? And does that in turn depend on which data constellations are available? How can users develop a good feeling for being on the right track when tuning? Answers to these questions are not only expected when it comes to optimally performing tuning per se, but also when it comes to making the tuning process transparent, i.e., answering the question why, after all, this and not that hyperparameter constellation was chosen.

This book delivers answers to the above questions, some of which were compiled as part of a study funded by the Federal Statistical Office of Germany. The contributed case studies and associated scripts also enable practitioners to reproduce the described tuning procedures and apply them themselves. The presented insights, cross-references, experiences, and recommendations will contribute to a better understanding of hyperparameter tuning in machine learning and to gain transparency.

Wiesbaden, Germany
March 2022

Florian Dumpert

Contents

Part II Applications

6 Hyperparameter Tuning and Optimization Applications 165
Thomas Bartz-Beielstein

7 Hyperparameter Tuning in German Official Statistics 177
Florian Dumpert and Elena Schmidt

8 Case Study I: Tuning Random Forest (Ranger) 187
Thomas Bartz-Beielstein, Sowmya Chandrasekaran,
Frederik Rehbach, and Martin Zaefferer

Contributors

Bartz Eva Bartz & Bartz GmbH, Gummersbach, Germany

Bartz-Beielstein Thomas Institute for Data Science, Engineering, and Analytics, TH Köln, Gummersbach, Cologne, Germany

Chandrasekaran Sowmya Institute for Data Science, Engineering, and Analytics, TH Köln, Gummersbach, Cologne, Germany

Dumpert Florian Federal Statistical Office of Germany, Wiesbaden, Germany

Mersmann Olaf Institute for Data Science, Engineering, and Analytics, TH Köln, Gummersbach, Germany

Rehbach Frederik Institute for Data Science, Engineering, and Analytics, TH Köln, Gummersbach, Cologne, Germany

Schmidt Elena Federal Statistical Office of Germany, Wiesbaden, Germany

Zaefferer Martin Bartz & Bartz GmbH and with Institute for Data Science, Engineering, and Analytics, TH Köln, Gummersbach, Germany;
Duale Hochschule Baden-Württemberg Ravensburg, Ravensburg, Germany

Abbreviations

ADAM	ADAptive Moment estimation algorithm
AI	Artificial Intelligence
APCS	Approximate Probability of Correct Selection
API	Application Programming Interface
AUC	Area Under the receiver operating characteristic Curve
AutoDL	Automated Deep Learning
AutoHAS	Automated Hyperparameter and Architecture Search
AutoML	Automated Machine Learning
BBOB	Black-Box Optimization Benchmarking
BCE	Binary Cross Entropy
BFGS	Broyden, Fletcher, Goldfarb, and Shanno
BO	Bayesian Optimization
BOHB	Bayesian Optimization HyperBand
CART	Classification and Regression Trees
CASH	Combined Algorithm Selection and Hyperparameter optimization
CDF	Cumulative Distribution Function
CFD	Computational Fluid Dynamics
CG	Conjugate Gradient
CID	Census-Income (KDD) Data Set
CIFAR-10	Canadian Institute for Advanced Research, 10 classes
CM	Consensus Method
CMA-ES	Covariance Matrix Adaptation Evolution Strategy
CNN	Convolutional Neural Network
CPPS	Cyber-physical Production Systems
CRAN	Comprehensive R Archive Network
CSV	Comma Separated Values
CV	Cross Validation
DACE	Design and Analysis of Computer Experiments
DDPG	Deep Deterministic Policy Gradient
DE	Differential Evolution

DeepOBS	Deep Learning Optimizer Benchmark Suite
DL	Deep Learning
DNN	Deep Neural Network
DOE	Design of Experiments
DT	Decision Tree
EA	Evolutionary Algorithm
EC	Evolutionary Computation
EDA	Exploratory Data Analysis
EGC	Elevator Group Control
EGO	Efficient Global Optimization
EI	Expected Improvement
EN	Elastic Net
ES	Evolution Strategy
GB	Gradient Boosting
GP	Gaussian Process
HB	Hyperband
HDF	Hierarchical Data Format
HPC	High Performance Computing
HPO	Hyperparameter Optimization
HPT	Hyperparameter Tuning
IIA	Independence of irrelevant alternatives
IID	Independent and Identically Distributed
ILS	Iterative Local Search
IOH	Iterative Optimization Heuristic
IOHanalyzer	Iterative Optimization Heuristics analyzer
IOHexperimenter	Iterative Optimization Heuristics experimenter
IOHprofiler	Iterative Optimization Heuristics profiler
IRACE	Iterative Racing
KNN	K-Nearest-Neighbor
LHD	Latin Hypercube Design
LHS	Latin Hypercube Sampling
LOOCV	Leave One Out Cross Validation
MAE	Mean Absolute Error
MAMP	Multiple Algorithm Multiple Problem
MASP	Multiple Algorithm Single Problem
MC	Monte Carlo
ML	Machine Learning
MLE	Maximum Likelihood Estimation
MMCE	Mean Mis-Classification Error
MOO	Multi Objective Optimization
MSE	Mean Squared Error
NACE	Nomenclature statistique des Activités économiques dans la Communauté Européenne
NADAM	Nesterov-accelerated Adaptive Moment Estimation
NAS	Neural Architecture Search

NIID	Normal, Independent and Identically Distributed
NLP	Natural Language Processing
NM	Nelder and Mead Simplex Algorithm
NN	Neural Network
OCBA	Optimal Computing Budget Allocation
OpenML	Open Machine Learning
PDF	Probability Distribution Function
PI	Probability of Improvement
R	R software environment for statistical computing and graphics
RF	Random Forest
RL	Reinforcement Learning
RMSE	Root Mean Squared Error
RMSProp	Root Mean Square Propagation
RNG	Random Number Generator
RNN	Recurrent Neural Network
RS	Random Search
RSM	Response Surface Methodology
SAMP	Single Algorithm Multiple Problem
SASP	Single Algorithm Single Problem
SGD	Stochastic Gradient Descent
SMAC	Sequential Model-Based Optimization for General Algorithm Configuration
SMBO	Surrogate Model Based Optimization
SPO	Sequential Parameter Optimization
SPOT	Sequential Parameter Optimization Toolbox
SPOTMisc	Sequential Parameter Optimization Toolbox—Miscelleanous Functions
SVM	Support Vector Machine
TF	TensorFlow
THK	Technische Hochschule Köln
TPE	Tree of Parzen Estimator
UCI	University of California, Irvine
XGBoost	Extreme Gradient Boosting

Chapter 1
Introduction

Eva Bartz

It's about resources, time, money, and effort. It's about how science serves our society.

Ever wondered how ice cream manufacturers picked the favorite flavor of the year? Might there be a "fruit conspiracy" (Sect. 5.2)?

The COVID-19 pandemic showed the importance of forecasts concerning healthcare workers, protective equipment, vaccines, and so forth. How do we model the data we have to generate sound results and robust conclusions?

How many places for childcare will we need in the future? How many teachers? Local politicians will need numbers to prepare good politics caring for our future, raising our children with the hope for a good education.

The needs and the possibilities for good forecasts and predictions are numerous in our society. Coming from the business end of things I was deeply impressed, how scientific research in general and hyperparameter tuning in particular changes and contributes to our society again and again.

Bartz & Bartz GmbH initiated the methods described in this book to achieve better results in hyperparameter tuning faster, with less effort and costs. Because, let's face it, computational time entails a number of costs. First and foremost it entails the time of the researcher, furthermore a lot of energy. All this equals money. So if we manage to achieve better results in hyperparameter tuning in less time, everybody profits. On a larger scale the methods described may contribute a small part to address some of the challenges we face as a society.

Having initiated the methods in an expertise funded by the Federal Statistical Office of Germany (destatis), we realized that a number of people and businesses might benefit from our knowledge. To be able to enlarge our entrepreneurial effort into a book, scientists from the Institute for Data Science, Engineering, and Analytics

E. Bartz (✉)
Bartz & Bartz GmbH, Gummersbach, Germany
e-mail: eva.bartz@bartzundbartz.de

© The Author(s) 2023
E. Bartz et al. (eds.), *Hyperparameter Tuning for Machine and Deep Learning with R*,
https://doi.org/10.1007/978-981-19-5170-1_1

1

of the Technische Hochschule Köln (THK) took over. We added the academic point of view to the business consulting, Bartz & Bartz GmbH provided.

Our clients from destatis contributed Chap. 7, discussing the "Hyperparameter Tuning in German Official Statistics".

We link academic and entrepreneurial requirements, hoping to create a very broad theoretical overview with high practical value for our readers. Thus this book can be used as a handbook as well as a text book. It provides hands-on examples that illustrate how hyperparameter tuning can be applied in practice and gives deep insights into the working mechanisms of Machine Learning (ML) and Deep Learning (DL) methods. Programming code is provided so that users can reproduce the results.

ML and DL methods are becoming more and more important and are used in many industrial production processes, e.g., Cyber-physical Production Systems (CPPS). Several hyperparameters of the methods used have to be set appropriately. Previous projects carried out produced inconsistent results in this regard. For example, with Support Vector Machines (SVMs) it could be observed that the tuning of the hyperparameters is critical to success with the same data material, with random forests the results do not differ too much from one another despite different selected hyperparameter values. While some methods have only one or a few hyperparameters, others provide a large number. In the latter case, optimization using a (more or less) fine grid (grid search) quickly becomes very time-consuming and can therefore no longer be implemented. In addition, the question of how the optimality of a selection can be measured in a statistically valid way (test problem: training/validation/test data and resampling methods) arises for both many and a few hyperparameters. In real-world projects, DL experts have gained profound knowledge over time as to what reasonable hyperparameters are, i.e., Hyper Parameter Tuning (HPT) skills are developed. These skills are based on human expert and domain knowledge and not on valid formal rules.

Figure 1.1 illustrates how data scientists select models, specify metrics, pre-process data, etc. Kedziora et al. (2020) present a similar description. Chollet and Allaire (2018) describe the situation as follows:

> If you want to get to the very limit of what can be achieved on a given task, you can't be content with arbitrary [hyperparameter] choices made by a fallible human. Your initial decisions are almost always suboptimal, even if you have good intuition. You can refine your choices by tweaking them by hand and retraining the model repeatedly—that's what machine-learning engineers and researchers spend most of their time doing.

> But it shouldn't be your job as a human to fiddle with hyperparameters all day—that is better left to a machine.

Please compare this to Fig. 2.2, which shows how the automated tuning process works. But is it reasonable to transfer the power or decision-making entirely to a machine? I don't think so and you probably don't either. But how do we accomplish that?

This book deals with the hyperparameter tuning of ML and DL algorithms and keeps the human in the loop. In particular, it provides

- a survey of important model parameters;

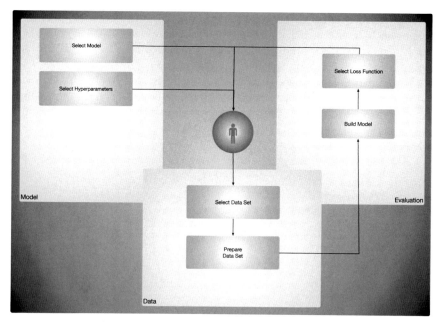

Fig. 1.1 Elements of the HPT process. For a given ML or DL model and its hyperparameters, the data scientist selects a data set and prepares the corresponding data. The model is built and a loss function is computed to evaluate the results

- three parameter tuning studies;
- one extensive global parameter tuning study;
- statistical analysis of the performances of the ML and DL methods based on severity; and
- a new way, based on consensus ranking, to analyze results from multiple algorithms.

More than 30 hyperparameters from six relevant ML methods and DL methods are analyzed. We extend the well-established SPOT framework that improves the optimization of ML methods and DL while increasing the transparency and keeping the human in the loop. The case studies presented in this book can be run on a regular desktop or notebook computer. No high-performance computing facilities are required. Interactive tools for visualization with the popular R package `ploty` are provided. We hope that you achieve better results with significantly less time, costs, effort, and resources using the methods described in this book. We wish you a successful implementation.

This book is structured as follows: Chap. 2 introduces the methodology. Chapter 3 presents models (algorithms or methods) and hyperparameters. HPT methods are introduced in Chap. 4. Chapter 5 discusses result aggregation and severity. Chapter 6 describes the relevance of HPT in industry. Chapter 7 presents HPT in official statistics. Four case studies are presented next. These HPT studies are using the Census-

Income (KDD) Data Set (CID), which will be described in Sect. 8.2.1. The first case study, which describes HPT for random forests, is presented in Chap. 8. This case study might serve as a starting point for the interested reader. The second case study analyzes Extreme Gradient Boosting (XGBoost) and is presented in Chap. 9. The third case study analyzes hyperparameter tuning for DL in Chap. 10. To expand on the example in Chap. 10, which considered tuning a Deep Neural Network (DNN), Chap. 11 also deals with neural networks, but focuses on a different type of learning task: Reinforcement Learning (RL). A global study, which analyzes *tunability*, is presented in Chap. 12.

Part I
Theory

Chapter 2
Tuning: Methodology

Thomas Bartz-Beielstein, Martin Zaefferer, and Olaf Mersmann

Abstract This chapter lays the groundwork and presents an introduction to the process of tuning Machine Learning (ML) and Deep Learning (DL) hyperparameters and the respective methodology used in this book. The key elements such as the hyperparameter tuning process and measures of tunability and performance are defined. Practical considerations are presented and all the ingredients needed for successful hyperparameter tuning are explained. A special focus lies on how to prepare the data. This might be the most thorough overview presented yet.

2.1 Introduction to Hyperparameter Tuning

This book deals with the tuning of the hyperparameters of methods from the field ML and DL, focusing on supervised learning (both regression and classification). In the following, the scope of this work is explained, including terminology of important terms (Tables 2.1 and 2.2).

The *data points* x come from an input data space $\mathcal{X}$ ($x \in \mathcal{X}$) and can have different scale levels (nominal: no order; ordinal: order, no distance; cardinal: order, distances). Nominal and ordinal data are mostly discrete, and cardinal data are continuous. Usually, the data points are k-dimensional vectors. The vector elements are also called *features* or independent variables. The number of data points in a data set is n.

T. Bartz-Beielstein (✉) · O. Mersmann
Institute for Data Science, Engineering and Analytics, TH Köln, Gummersbach, Germany
e-mail: thomas.bartz-beielstein@th-koeln.de

O. Mersmann
e-mail: olaf.mersmann@th-koeln.de

M. Zaefferer
Bartz & Bartz GmbH and with Institute for Data Science, Engineering, and Analytics, TH Köln, Gummersbach, Germany
e-mail: zaefferer@dhbw-ravensburg.de

Duale Hochschule Baden-Württemberg Ravensburg, Ravensburg, Germany

E. Bartz et al. (eds.), *Hyperparameter Tuning for Machine and Deep Learning with R*,
https://doi.org/10.1007/978-981-19-5170-1_2

Table 2.1 Symbols used in this book

Symbol	Name and description
$\mathcal{A}$	Algorithm, model, methods (see Definition 2.20)
$A_{\lambda(t)}$	Model with hyperparameter configuration λ at time step t
$A_{\lambda(*)}$	Model with best hyperparameter configuration λ
c	Number of classes/categories
$c_{1-\alpha}$	Threshold value or the cut-off point (used in hypothesis testing)
$d(\cdot)$	Test statistic
d_j	Difference between the j-th samples
$E()$	Expectation
$\mathrm{Err}_{\mathrm{test}}$	Test error, generalization error
f	Function that describes relationship between input and output
$f_{\mathrm{acc}}^{(\mathrm{train})}$	Training accuracy
$f_{\mathrm{acc}}^{(\mathrm{test})}$	Test accuracy
$f_{\mathrm{acc}}^{(\mathrm{val})}$	Validation accuracy
H_0	Null hypothesis
H_1	Alternative hypothesis
$\mathbf{I}$	Indicator function
k	(Problem) dimension
$k(x, x')$	Kernel function
$\mathcal{L}$	Loss function
n	Sample size or number of observations
n_{init}	Initial design size
N	Total number of samples
N_{Feats}	Number of features
$\mathbb{N}$	Natural numbers
O	Optimizer
p	Norm, e.g., used in Eq. (2.1) or number of model coefficients
$p(\mathcal{A})$	Performance of algorithm $\mathcal{A}$
$P_H(x)$	Probability of x under the hypothesis H
$\mathbb{R}$	Set of real numbers
S_d	Sample standard deviation of differences
S_{nr}	Severity (non-rejection)
S_r	Severity (rejection)
S	Surrogate (model)
T_i	i-th tree
$|T|$	Number of splits for a tree
$\mathcal{T}$	Tuner
$u_{1-\alpha}$	Upper $1 - \alpha$ quantile of the normal distribution
$\mathcal{X}$	Input data space or complete data set

(continued)

Table 2.1 (continued)

Symbol	Name and description
$((\mathcal{X}, \mathcal{Y})_{\text{CID}}$	The full Census Income Data (CID) data set ($n = 299\,285$ observations)
$(X, Y)^{(\text{test})}$	Test data set
$(X, Y)^{(\text{train})}$	Training data set
$(X, Y)^{(\text{train} \cup \text{val})}$	Training and validation data set
$(X, Y)^{(\text{valtrain})}$	Validation set, subset of $(X, Y)^{(\text{train})}$, internally used by the model $\mathcal{A}$ in the inner optimization loop
$(X, Y)^{(\text{val})}$	Validation data set
t	Iteration counter. I.g., the t-th SPOT surrogate will be denoted as $\mathcal{S}(t)$
T	Transpose of a matrix or vector
x	Data point, feature, predictor, covariates, explanatory variable
$\overline{x}$	Observed difference
x^*	New/unknown data point
X	Data, usually partitioned into training, validation, and test data
y	Actually observed value, outcome, response variable, label
$y_{\text{val}}^{(*)}$	Best function value on the validation data $(X, Y)^{(\text{val})}$
$y_{\text{val}}^{(\text{OCBA*})}$	Best function value on the validation data $(X, Y)^{(\text{val})}$, evaluated with OCBA
$\hat{y}_i$	Predicted value for the i-th observed value y_i
y^*	Output value for a new/unknown data point
$\overline{y}$	Mean difference between two observations, usually for paired samples
$\mathcal{Y}$	Output data space (dependent variables)

In addition, we consider output data (dependent variables) $y \in \mathcal{Y}$. These can also have different scale levels (nominal, ordinal, cardinal). Output data are usually scalar. Variable identifiers like x and y represent scalar or vector quantities, the meaning can be deduced from the context. In many practical applications, e.g., in the case studies in this book, a subset, (X, Y), of the full data set $(\mathcal{X}, \mathcal{Y})$ is used.

Definition 2.1 (*Supervised learning*) *Supervised learning* is a branch of ML where models are trained on *labeled* data sets to make predictions.

For each observation, both the input $x_i \in \mathcal{X}$ and the outcome $y_i \in \mathcal{Y}$ ($i = 1, 2, \ldots, n$) is known during training. A supervised learning algorithm $\mathcal{A}$ learns the relationship between the inputs and output. That is, given a training data set (x_i, y_i), $i = 1, \ldots, n$, it returns a *model* $f : \mathcal{X} \mapsto \mathcal{Y}$ with which we can predict the expected value of y^* given x^*:

$$\hat{y^*} = f(x^*)$$

Table 2.2 Greek Symbols used in this book

Symbol	Name	Comment, Example
α	Significance level (in hypothesis testing)	Probability of a type I error, given that the null hypothesis is true
β	Regression coefficient; the probability of a type II error, given that the alternative hypothesis is true	
Δ	Relevant difference	
μ	Mean of a random variable	
λ	Hyperparameter configuration	Also: nugget in Kriging
λ_i	i-th hyperparameter configuration	Used in SMBO
λ_0	Default hyperparameter configuration	Used in SMBO
$\lambda^\star$	Best hyperparameter configuration	Best configuration in theory
$\hat{\lambda}$	Best hyperparameter configuration obtained by evaluating a finite set of samples	Best configuration "in practice"
Λ	Hyperparameter space	
π	Problem instance	
Φ	Cumulative distribution function of the standard normal distribution	
Ψ	Hyperparameter response space	
ψ_i	Hyperparameter response surface function evaluated for the i-th hyperparameter configuration λ_i	
$\psi^{(test)}$	Hyperparameter response surface function (on test data)	As defined in Eq. (2.11)
$\psi^{(train)}$	Hyperparameter response surface function (on train data)	
$\psi^{(val)}$	Hyperparameter response surface function (on validation data)	As defined in Eq. (2.9)
σ^2	Variance	
τ	Possible values (in hypothesis testing)	
θ	Set of parameters	Also: Kriging hyperparameters

Definition 2.2 (*Regression*) If a supervised learning problem has an infinite number of labels or outcomes, i.e., $\mathcal{Y} \subseteq \mathbb{R}$, then it is called a *regression problem* and the task of finding a model that captures the relationship between the input space and output space is called *regression*.

Definition 2.3 (*Classification*) If a supervised learning problem has a finite number of labels or outcomes, i.e., $\mathcal{Y} = a_1, a_2, \ldots, a_c$ with $c \in \mathbb{N}$, it is called a *classification* *problem* and the task of finding a model that captures the relationship between the input space and output space is called *classification*.

Example: Regression and Classification

A typical question in regression is "how does relative humidity depend on temperature?", whereas a typical question in classification is "how does the default on a loan (yes, no) depend on the income of the borrower?"

We study ML and DL *algorithms*, also referred to as *models* or *methods*.

 Note

The terms "algorithm", "model", and "method" will be used interchangeably in this book. Their specific meanings can be derived from the context.

The learning algorithm itself has parameters called *hyperparameters*. Hyperparameters are distinct from model parameters.

Definition 2.4 (*Hyperparameter*) *Hyperparameters* are settings or configurations of the methods (models), which are freely selectable within a certain range and influence model performance (quality). One specific set of hyperparameters is denoted as λ, where Λ is the hyperparameter space.

Definition 2.5 (*Model parameters*) Model parameters are chosen during the learning process by the model itself.

Example: Hyperparameters and Model Parameters

The weights of the connections in an Neural Network (NN) are an example of model parameters, whereas the number of units or layers of a NN is a hyperparameter.

2.2   Performance Measures for Hyperparameter Tuning

2.2.1   Metrics

Metrics are used to measure the distance between points and then define the similarity between them.

Definition 2.6 (*Metric*) Let X denote a set. A *metric* is a function

$$d : X \times X \rightarrow \mathbb{R}_0^+$$

that returns the distance between any two values from the set X. Metrics are symmetric, positive definite, and fulfill the triangle equality.

The *Minkowski* distance, which will be used in this book, is defined as follows.

Definition 2.7 (*Minkowski Distance*)

$$d_p(x, x') = \left\| x - x' \right\|_p = \sqrt[p]{\sum_{i=1}^{n} \left| x_i - x_i' \right|^p}. \tag{2.1}$$

Some well-known metrics are special cases of the Minkowski distance:

- for $p = 1$ we get the *Manhattan distance*,
- for $p = 2$ we get the *Euclidean distance*,
- and for $p = \infty$ we get the *Chebyshev distance*.

2.2.2 Performance Measures

Several measures are used to evaluate the performance of ML and DL methods, because performance can be expressed in many different ways, e.g., as a measure of the fit of the model to the observed data values. The performance measure is evaluated after a single ML or DL training step and returns a value to assess the quality of the model or the prediction.

Tips: Measures in R

Basic metrics are implemented in the package SPOTMisc. The mlr tutorial "Implemented Performance Measures"[1] presents a comprehensible overview. The R package Metrics is also a valuable tool.

Before we can define performance measures for classification or regression problems, we need some tools from which we will build these up.

Definition 2.8 (*Loss function*) A function

$$\mathcal{L} : \mathcal{Y} \times \mathcal{Y} \rightarrow \mathbb{R}_0^+$$

---

[1] Available on https://mlr.mlr-org.com/articles/tutorial/measures.html.

is called a *loss function* or short a *loss*. $\mathcal{L}(y, \hat{y})$ is a measure of how "bad" it is to predict $\hat{y}$ given that the true label is y.

Concrete examples of loss functions are

Definition 2.9 (*Quadratic loss*) The loss function

$$\mathcal{L}_2(y, \hat{y}) = (y - \hat{y})^2$$

is called the *quadratic loss*.

Definition 2.10 (*Absolute value or L1 loss function*) The $L1$ or *absolute value loss* is defined as

$$\mathcal{L}_1(y, \hat{y}) = |y - \hat{y}|.$$

Definition 2.11 (*0-1 loss*) The loss function

$$\mathcal{L}_{01}(y, \hat{y}) = \mathbf{I}(y \neq \hat{y}) = \begin{cases} 0 & \text{when } y = \hat{y} \\ 1 & \text{else} \end{cases}$$

is called the *0-1 loss*.

Definition 2.12 (*Cross-entropy loss*) The *cross-entropy* loss function is defined as

$$\mathcal{L}_{CE}(y, \hat{y}) = y \log(\hat{y}) + (1 - y) \log(1 - \hat{y}) = \begin{cases} \log(\hat{y}) & \text{when } y = 1 \\ \log(1 - \hat{y}) & \text{when } y = 0 \end{cases}.$$

Note that it is only defined for binary outcomes y, while the predicted label $\hat{y}$ can be any value and is usually assumed to be a class probability.

With these loss functions, we can define *performance measures*. A performance measure evaluates the loss on a data set and returns an aggregate loss. Depending on the ML task, e.g., classification or regression, different categories of measures are useful. In the following, $\hat{y}_i = f(x_i)$ is the predicted value of the corresponding learning model $\mathcal{A}$ for the i-th observation, x_i is the i-th data point, and y_i is the actually observed value.

2.2.3 Measures for Classification

Definition 2.13 (*Mean Mis-Classification Error*) Mean Mis-Classification Error (MMCE) is defined as

$$\text{MMCE} = \frac{1}{n} \sum_{i=1}^{n} \mathcal{L}_{01}(y_i, \hat{y}_i) = \frac{1}{n} \sum_{i=1}^{n} \mathbf{I}(y_i \neq \hat{y}_i). \tag{2.2}$$

MMCE is used to evaluate the performance of the ML methods in the case studies (Chaps. 8–10) and in the global study (Chap. 12).

Definition 2.14 (*Binary cross-entropy loss*) Binary Cross Entropy (BCE) loss or *log-loss* is defined as

$$BCE = -\frac{1}{n} \sum_{i=1}^{n} \mathcal{L}_{CE}(y_i, \hat{y}_i) \tag{2.3}$$

The DL methods in Chap. 10 are tuned based on the $\psi^{(val)}$ (`validationLoss`), which computes the BCE loss.

Definition 2.15 (*Accuracy*) The *accuracy* is defined as

$$ACC = \frac{1}{n} \sum_{i=1}^{n} 1 - \mathcal{L}_{01}(y_i, \hat{y}_i) = \frac{1}{n} \sum_{i=1}^{n} \mathbb{I}(y_i = \hat{y}_i).$$

Note that, in contrast to the previous measures, higher accuracies are better than lower ones.

2.2.4 Measures for Regression

Typical regression measures are as follows.

Definition 2.16 (*Mean Squared Error*) The Mean Squared Error (MSE) is defined as

$$MSE = \frac{1}{n} \sum_{i=1}^{n} \mathcal{L}_2(y_i, \hat{y}_i).$$

When used for classification, the MSE is sometimes called the *Brier score*.

Definition 2.17 (*Root Mean Squared Error*) The Root Mean Squared Error (RMSE) is defined as the square root of the MSE:

$$RMSE = \sqrt{MSE} = \sqrt{\frac{1}{n} \sum_{i=1}^{n} \mathcal{L}_2(y_i, \hat{y}_i)}. \tag{2.4}$$

RMSE is used to evaluate the performance of the ML methods in the global study (Chap. 12).

Definition 2.18 (*Mean Absolute Error*) The Mean Absolute Error (MAE) is defined as

$$MAE = \frac{1}{n} \sum_{i=1}^{n} \mathcal{L}_1(y_i, \hat{y}_i).$$

2.3   Hyperparameter Tuning

After specifying the data, methods for supervised learning with their hyperparameters and performance measures, the hyperparameter tuning problem can be defined.

Definition 2.19 (*Hyperparameter tuning*) The determination of the best possible hyperparameters is called *tuning* (Hyperparameter Tuning (HPT)). HPT develops tools to explore the space of possible hyperparameter configurations *systematically*, in a structured way, i.e., HPT is an optimization problem.

The terms HPT and Hyperparameter Optimization (HPO) are often used synonymously. In the context of the analyses presented in this book, these terms have different meanings:

HPO   develops and applies methods to determine the best hyperparameters in an effective and efficient manner.

HPT   develops and applies methods that try to analyze the effects and interactions of hyperparameters to enable *learning and understanding*.

HPT can be seen as an extension of HPO, because it provides additional tools and keeps experimenters and applicants in the loop. The relationship between HPT and HPO can be formulated as follows:

$$HPO \subset HPT.$$

It simplifies the notation in the book: whenever HPT is mentioned, HPO is covered as well.

In a data-rich situation, the best HPT approach is to randomly partition the data set (X, Y) into three parts as illustrated in Fig. 2.1.

The following definitions are based on Hastie (2009) and Bergstra and Bengio (2012). The objective of a learning algorithm $\mathcal{A}$ is to find a function f that minimizes some expected loss $\mathcal{L}(y, f(x))$ over samples $(x, y) \in (X, Y)$.

Definition 2.20 (*Learning algorithm*) A *learning algorithm* $\mathcal{A}$ is a functional that maps a data set $(X, Y)^{(\text{train})}$ (a finite set of samples) to a function $f : X \to Y$, i.e., $\mathcal{A}((X, Y)^{(\text{train})}) \mapsto f$.

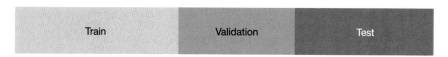

Train	Validation	Test

Fig. 2.1 Dataset split into three parts: (i) a training set $(X, Y)^{(\text{train})}$ used to fit the models, (ii) a validation set $(X, Y)^{(\text{val})}$ to estimate prediction error for model selection, and (iii) a test set $(X, Y)^{(\text{test})}$ used for assessment of the generalization error

A learning algorithm $\mathcal{A}$ can estimate f through the optimization of a training criterion with respect to a set of parameters, $\lambda \in \Lambda$. Because the true relationship f is unknown in real-world settings, $\mathcal{A}$ will return an estimation of f, which will be denoted as $\hat{f}$.

The learning algorithm itself often has hyperparameters $\lambda \in \Lambda$, and the actual learning algorithm is the one obtained after choosing λ, which can be denoted $\mathcal{A}_\lambda$.

The computation performed by $\mathcal{A}$ itself often involves an "inner optimization" problem, e.g., optimizing the weights of a NN. The hyperparameter optimization can be considered as an "outer-loop" optimization problem. It can be formulated as follows:

$$\lambda^{(*)} = \arg\min_{\lambda \in \Lambda} E_{(x,y) \in (\mathcal{X}, \mathcal{Y})} \left[\mathcal{L} \left(y, \mathcal{A}_\lambda ((X, Y)^{(\text{train})}) \right) \right]. \tag{2.5}$$

In practice, the underlying space $(\mathcal{X}, \mathcal{Y})$ is too large or the true relation between $\mathcal{X}$ and $\mathcal{Y}$ is unknown. Therefore, the validation set is used and Eq. (2.5) is replaced by

$$\lambda^{(*)} \approx \arg\min_{\lambda \in \Lambda} \frac{1}{\left| (X, Y)^{(\text{val})} \right|} \sum_{x \in (X, Y)^{(\text{val})}} \mathcal{L} \left(y, \mathcal{A}_\lambda ((X, Y)^{(\text{train})})(x) \right) \tag{2.6}$$

Practitioners are interested in a way to choose λ so as to minimize generalization or test error, which is based on unknown data to avoid overfitting. The generalization error can be defined as follows.

Definition 2.21 (*Generalization error (Test error)*) *Generalization error*, also referred to as *test error*, is the estimated loss over an independent (test) sample $(x, y) \in (X, Y)^{(\text{test})}$:

$$\text{Err}_{\text{test}} = E_{(x,y) \in (X,Y)^{(\text{test})}} \left[\mathcal{L}(y, \mathcal{A}_\lambda ((X, Y)^{(\text{train})})) \right].$$

Definition 2.22 (*Hyperparameter optimization problem*) The *hyperparameter optimization problem* can be stated in terms of a hyperparameter response function, $\psi \in \Psi$, as follows:

$$\lambda^{(*)} \approx \arg\min_{\lambda \in \Lambda} \psi(\lambda) \approx \arg\min_{\{\lambda^{(i)}\}_{i=1,2,\dots,n}} \psi(\lambda) = \hat{\lambda}, \tag{2.7}$$

where

$$\psi(\lambda)^{(\text{test})} = \frac{1}{|(X, Y)^{(\text{test})}|} \sum_{x \in (X,Y)^{(\text{test})}} \mathcal{L} \left(y, \mathcal{A}_\lambda ((X, Y)^{(\text{train})}) \right). \tag{2.8}$$

! Attention: Validation and Test Data

The validation set $(X, Y)^{(\text{val})}$ is used during optimization to estimate the prediction error for model selection,

$$\psi(\lambda)^{(\text{val})} = \frac{1}{|(X, Y)^{(\text{val})}|} \sum_{x \in (X,Y)^{(\text{val})}} \mathcal{L}\left(y, \mathcal{A}_\lambda((X, Y)^{(\text{train})})\right), \qquad (2.9)$$

whereas the test set in Eq. (2.8) is used for the assessment of the generalization error of the selected model.

Summarizing, we can define HPO in ML and DL as a minimization problem

Definition 2.23 (*Hyperparameter optimization*) Hyperparameter optimization is the minimization of

$$\psi(\lambda) \text{ over } \lambda \in \Lambda.$$

Definition 2.24 (*Hyperparameter surface*) Similar to the definition in Design of Experiments (DOE), the function $\psi \in \Psi$ is referred to as the hyperparameter *response surface*.

Different data sets, tasks (classification or regression), and methods define different sets Λ and functions Ψ.

A natural strategy for finding an adequate λ is described in Eq. (2.7): a set of candidate solutions, $\{\lambda^{(i)}\}_{i=1,2,\ldots,n}$, is chosen. Then $\psi(\lambda)$ is computed for each one, and the best hyperparameter configuration is returned as $\tilde{\lambda}$.

Whereas λ denotes an arbitrarily chosen hyperparameter configuration, important hyperparameter configurations will be labeled as follows: λ_i is the i-th hyperparameter configuration, λ_0 is the initial hyperparameter configuration, $\lambda^{(*)}(t)$ is the best hyperparameter configuration at iteration t, and $\lambda^{(*)}$ is the final best hyperparameter configuration.

Definition 2.25 (*Low effective dimension*) If a function f of two variables could be approximated by another function of one variable ($f(x_1, x_2) \approx g(x_1)$), we could say that f has a *low effective dimension*.

Several approaches exist for the tuning procedure. A model-based search is presented in this book. The corresponding model is called a surrogate model, or *surrogate*, S, for short.

In Sect. 5.8, different experimental designs for benchmarking optimization methods are discussed: the most simple design evaluates *one* single algorithm on *one* problem, whereas the most complex design is used for comparing multiple algorithms on multiple problems. HPT can be seen as a variant of the simple design, because the experimenter is interested in the improved performance of one method on one problem. To obtain this goal, the best hyperparameter configuration is determined. However, this simple setting can be extended, because the performance of the tuned method can be compared to the performance of some default method or to a competitive state-of-the art method. In the latter case, the hyperparameters of the state-of-the-art method should also be tuned to enable a fair comparison. Although HPT is not a benchmarking method on its own, it can be seen as a prerequisite for a fair and sound benchmark study. Note that the complex design requires adequate statistical methods for the comparison. An approach based on consensus ranking is presented in Chap. 12.

Tips: How to Select a Performance Measure

Kedziora et al. (2020) state that in classification "unsurprisingly", *accuracy*[2] is considered the most important performance measure. Accuracy might be an adequate performance measure for classification of balanced data. For unbalanced data, other measures are better. In general, there are many other ways to measure model quality, e.g., metrics based on time complexity and robustness or the model complexity (interpretability, see also Definition 2.27) Bartz-Beielstein et al. (2020a).

In contrast to classical optimization, where the same optimization function can be used for tuning and final evaluation, training of MLs and DL methods faces a different situation: Training and validation are usually based on the loss function whereas the final evaluation is based on a different measure, e.g., accuracy.

It is important to distinguish between estimates of performance (minimization of the generalization error) based on validation and test sets. The loss function, which has desirable mathematical properties, e.g., differentiability, acts as a surrogate for the performance measure the user is finally interested in. Several performance measures are used at different stages of the HPT procedures:

1. training loss, i.e., $\psi^{(train)}$,
2. training accuracy, i.e., $f_{acc}^{(train)}$,
3. validation loss, i.e., $\psi^{(val)}$,
4. validation accuracy, i.e., $f_{acc}^{(val)}$,
5. test loss, i.e., $\psi^{(test)}$, and
6. test accuracy, i.e., $f_{acc}^{(test)}$.

This complexity gives reason for the following question:

Question: Which performance measure should be used during the HPT (HPO) procedure?

Most authors recommend using test accuracy or test loss as the measure for hyperparameter tuning Schneider et al. (2019). In order to understand the correct usage of these performance measures, it is important to look at the goals, i.e., selection or assessment, of a tuning study.

To keep the discussion focus, accuracy was used in the previous considerations. Instead of accuracy, other measures, e.g., MMCE, can be considered.

---

[2] Accuracy in binary classification is the proportion of correct predictions among the total number of observations Metz (1978).

2.4   Model Selection and Assessment

Hastie et al. (2017) state that selection and assessment are two separate goals:

Selection: Estimating the performance of different models in order to choose the best one. Model selection is important *during* the tuning procedure.

Assessment: Model assessment is used for the *final* report (evaluation of the results). Having chosen a final hyperparameter configuration, $\lambda^{(*)}$, the assessment estimates the model's prediction error (generalization error) on new data based on Eq. (2.8). This determines whether predicted values from the model are likely to accurately predict responses on future observations or samples from the hold-out set $(X, Y)^{(test)}$. This process may help to prevent problems such as overfitting.

In principle, there are two ways of model assessment and selection Hastie et al. (2017):

1. External assessment/selection uses different sets of data. The first p data samples are for model training and $n - p$ for validation. An explicit hold-out data set is used. Problem: holding back data from model fitting results in lower precision and power.
2. Internal assessment/selection uses data splitting and resampling methods. The true error might be *underestimated*, because the same data samples that were used for fitting the model are used for prediction. The so-called in-sample (also apparent, or resubstitution) error is smaller than the true error.

The test set $(X, Y)^{(test)}$ should be used only at the end of the HPT procedure. It should not be used during the training and validation phase, because if the test set is used repeatedly, e.g., for choosing the model with smallest test-set error, "the test set error of the final chosen model will underestimate the true test error, sometimes substantially." Hastie et al. (2017).

The following example shows that there is no general agreement on how to use training, validation, and test sets as well as the associated performance measures.

Example: Basic Comparisons in Manual Search

Wilson et al. (2017) describe a manual search. They allocated a pre-specified budget on the number of epochs used for training each model.

- When a test set was available, it was used to chose the settings that achieved the best peak performance on the test set by the end of the fixed epoch budget.
- If no explicit test set was available, e.g., for Canadian Institute for Advanced Research, 10 classes (CIFAR-10), they chose the settings that achieved the lowest training loss at the end of the fixed epoch budget.

Theoretically, the results from the internal assessment are not of interest because new data values are not likely to coincide with their training set values. Bergstra and Bengio (2012) stated that "because of finite data sets, test error is not monotone in validation error, and depending on the set of particular hyperparameter values λ evaluated, the test error of the best-validation error configuration may vary", i.e.,

$$\psi_i^{(\text{val})} < \psi_j^{(\text{val})} \implies \psi_i^{(\text{test})} < \psi_j^{(\text{test})},$$

where $\psi_i^{(\cdot)}$ denotes the value of the hyperparameter response surface for the i-th hyperparameter configuration λ_i.

Furthermore, the estimator, e.g., for loss, obtained by using a single hold-out test set usually has high variance. Therefore, Cross Validation (CV) methods were proposed. Hastie et al. (2017) concluded

> that estimation of test error for a particular training set is not easy in general, given just the data from that same training set. Instead, cross-validation and related methods may provide reasonable estimates of the expected error.

The standard practice for evaluating a model found by CV is to report the hyperparameter configuration that minimizes the loss on the validation data, i.e., $\hat{\lambda}$ as defined in Eq. (2.7). Repeated CV, i.e., k-fold CV, reduces the variance of the estimator and results in a more accurate estimate. There is, as always, a trade-off: the more CV folds, the better the estimate, but more computational time is needed.

Example: Reporting the model assessment (final evaluation)

It can be useful to take the uncertainty due to the choice of hyperparameters values into account, when one reports the performance of learning algorithms. Bergstra and Bengio (2012) present a procedure for estimating test set accuracy, which takes into account any uncertainty in the choice of which trial is actually the best-performing one. To explain this procedure, they distinguish between estimates of performance $\psi^{(\text{val})}$ and $\psi^{(\text{test})}$ based on the validation and test sets, respectively.

To resolve the difficulty of choosing the best configuration, Bergstra and Bengio (2012) reported a weighted average of all the test set scores, in which each one is weighted by the probability that its particular λ_s is in fact the best. In this view, the uncertainty arising from $X^{(\text{val})}$ being a finite sample makes the test-set score of the best model among $\{\lambda_i\}_{i=1,2,\dots,n}$ a random variable.

2.5 Tunability and Complexity

The term *tunability* is used according to the definition presented in Probst et al. (2019a).

Definition 2.26 (*Tunability*) Tunability describes a measure for modeling algorithms as well as for individual hyperparameters. It is the difference between the model quality for default values (or reference values) and the model quality for optimized values (after tuning is completed).

Or in the words of Probst et al. (2019a): "measures for quantifying the tunability of the whole algorithm and specific hyperparameters based on the differences between the performance of default hyperparameters and the performance of the hyperparameters when this hyperparameter is set to an optimal value". Tunability of individual hyperparameters can also be used as a measure of their *relevance, importance,* or *sensitivity.* Accordingly, parameters with high tunability are of greater importance for the model. The model reacts strongly to (i.e., is sensitive to) changes in these hyperparameters.

Definition 2.27 (*Complexity*) The term *complexity* or model complexity generally describes, how many functions of different difficulty can be represented by a model.

Example: Complexity

For linear models, complexity can be influenced by the number of model coefficients. For Support Vector Machines (SVMs), it can be influenced by the parameter `cost`.

2.6   The Basic HPT Process

Now all ingredients are available for defining the basic HPT process.

Definition 2.28 (*The basic HPT process*) For a given space of hyperparameters Λ, a ML or DL model $\mathcal{A}$ with hyperparameters λ, training, validation, and testing data $(X, Y)^{(\text{train})}$, $(X, Y)^{(\text{val})}$, and $(X, Y)^{(\text{test})}$, respectively, a loss function $\mathcal{L}$, and a hyperparameter response surface function ψ, e.g., mean loss, the basic HPT process looks like this

(HPT-1) Set $t = 1$. Hyperparameter selection (at iteration t). Choose a set of hyperparameters from the space of hyperparameters, $\lambda(t) \in \Lambda$.

(HPT-2) ML or DL model building. Build the corresponding ML or DL model $\mathcal{A}_{\lambda(t)}$. Note: The model building step is listed separately, because this corresponds with the steps of the `keras` procedure: first the DL model is specified and compiled (via `compile`), then it is trained (e.g., via `fit`).

(HPT-3) ML or DL model training and evaluation (e.g., via `keras fit`). Fit the model $\mathcal{A}_{\lambda(t)}$ to the training data $(X, Y)^{(\text{train})}$ (see Fig. 2.1) and measure the final performance, e.g., expected loss, on the validation data $(X, Y)^{(\text{val})}$, see Eq. (2.9). Under k-fold CV, the performance measure from Eq. (2.9) can be written as

$$\psi_{\text{CV}}^{(\text{val})} = \frac{1}{k} \sum_{i=1}^{k} \frac{1}{|(X,Y)^{(\text{val})}|} \sum_{x \in (X,Y)_i^{(\text{val})}} \mathcal{L}\left(y, \mathcal{A}_{\lambda^{(t)}}((X,Y)_i^{(\text{train})})\right), \qquad (2.10)$$

if the training and validation set partitions are generated k times.

(HPT-4) Hyperparameter update. The next set of hyperparameters to try, $\lambda(t+1)$, is chosen accordingly to minimize the performance, e.g., $\psi^{(\text{val})}$. An infill criterion (acquisition function) is used.

(HPT-5) Looping. Repeat until budget is exhausted.

(HPT-6) Final evaluation of the best hyperparameter set $\lambda^{\star}$ on test (or hold out) data $(X,Y)^{(\text{test})}$, i.e., measuring performance on the test data

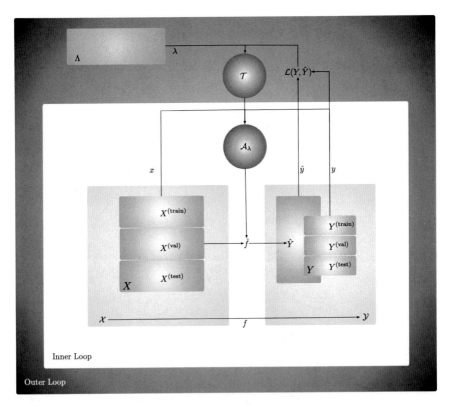

Fig. 2.2 Elements of the HPT process. A *dark gray* background color shows the "outer" optimization loop, whereas a *light gray* background is used for the "inner" optimization loop. $\mathcal{T}$ denotes the tuner, which selects hyperparameters $\lambda \in \Lambda$ to optimize the loss $\mathcal{L}$. The loss is the output from the inner optimization loop, where an algorithm $\mathcal{A}_{\lambda}$ (also referred to as the model or the method) optimizes the function $\hat{f}$, that builds a relation between input data $X \in \mathcal{X}$ and output data $Y \in \mathcal{Y}$. Note, depending on the three different tasks, i.e., model training, validation (selection), or test (assessment), the loss function $\mathcal{L}$, computes tree different losses: $\psi^{(\text{train})}$, $\psi^{(\text{val})}$, and $\psi^{(\text{test})}$

$$\psi^{(\text{test})} = \frac{1}{|(X, Y)^{(\text{test})}|} \sum_{x \in (X,Y)^{(\text{test})}} \mathcal{L}\left(y, \mathcal{A}_{\lambda^{(*)}}((X, Y)^{(\text{train} \cup \text{val})})\right). \quad (2.11)$$

The HPT process is illustrated in Fig. 2.2.

Essential for this process is the infill criterion (acquisition function) in (HPT-4). It uses the validation performance to determine the next set of hyperparameters to evaluate. and requires building and training a new model. Often, the hyperparameter space Λ is not differentiable or even continuous. Gradient methods are not applicable in Λ. Pattern search, Evolution Strategys (ESs), or other gradient-free methods are used instead.

2.7 Practical Considerations

Unfortunately, training, validation, and test data are used inconsistently in HPO studies: for example, Wilson et al. (2017) selected *training loss*, $\psi^{(\text{train})}$, (and not validation loss) during optimization and reported results on the test set $\psi^{(\text{test})}$.

Choi et al. (2019) considered this combination as a "somewhat non-standard choice" and performed tuning (optimization) on the validation set, i.e., they used $\psi^{(\text{val})}$ for tuning, and reported results $\psi^{(\text{test})}$ on the test set. Their study allows some valuable insight into the relationship of validation and test error:

> For a *relative comparison* between models during the tuning procedure, in-sample error is convenient and often leads to effective model selection. The reason is that the relative (rather than absolute performance) error is required for the comparisons. Choi et al. (2019)

Choi et al. (2019) compared the final predictive performance of NN optimizers after tuning the hyperparameters to minimize validation error. They concluded that their "final results hold regardless of whether they compare final validation error, i.e., $\psi^{(\text{val})}$, or test error, i.e., $\psi^{(\text{test})}$". Figure 1 in Choi et al. (2019) illustrates that the relative performance of optimizers stays the same, regardless of whether the validation or the test error is used. Choi et al. (2019) considered two statistics: (i) the quality of the best solution and (ii) the speed of training, i.e., the number of steps required to reach a fixed validation target.

2.7.1 Some Thoughts on Cross Validation

There are some drawbacks of k-fold CV: at first, the choice of the number of observations to be held out from each fit is unclear: if n denotes the size of the training data set, with $k = n$, which is referred to as Leave One Out Cross Validation (LOOCV), the CV estimator is approximately unbiased for the expected prediction error. But this estimator has high variance, because LOOCV does not mix the observations very much. The estimates from each fold are highly correlated and hence their average can have high variance. Furthermore, computational costs are relatively high, because n evaluations of the model are necessary.

Furthermore, CV does not fully represent variability of variable selection, because p elements are removed each time from set of n. Kohavi (1995) reviewed accuracy estimation methods and compared CV and bootstrap Efron and Tibshirani (1993). Note that Picard and Cook (1984) proposed Monte Carlo (MC) CV as an improvement over standard CV.

2.7.2 Replicability and Stochasticity

Results from DL and ML tuning runs are noisy, e.g., caused by random sampling of batches and initial parameters. Repeats to estimate means and variances that are required for a sound statistical analysis are costly.

However, even if seeds are provided, full reproducibility cannot be guaranteed. Gramacy (2020) mentioned two important issues:

- First, Random Number Generator (RNG) sequences can vary across software versions.
- Second, conditional expressions involving floating point calculations can change across hardware architectures and lead to different results in stochastic experimentation even with identical pseudorandom numbers.

As a consequence, it is impossible to fully remove randomness from the experiments. López-Ibáñez et al. (2021b) provide guidelines and suggest tools that may help to overcome some of these reproducibility issues.

2.7.3 Implementation in R

Background: Data Types in R

Our implementation is done in the R programming language, where data and functions are represented as objects. Each object has a data type. The basic (or *atomic*) data types are shown in Table 2.3.

In addition to these data types, R uses an *internal storage mode* which can be queried using `typeof()`. Thus, there are two storage modes for the `numeric` data type:

- `integer` for integers and
- `double` for real values.

The corresponding variables are referred to as *numeric*.

Table 2.3 Atomic data types of the programming language R

Data type	Description	Examples
NULL	Empty set	NULL
logical	Boolean values	TRUE, FALSE
numeric	Integer and real values	1, 0.5
complex	Complex values	1+1i
character	Characters and strings of characters	"123", "test"

Factors are used in R to represent nominal (qualitative) features. Ordinal features can also be represented by factors in R (see argument `ordered` of the function `factor()`). However, this case is not considered here. Factors are generated with the generating function `factor()`. Factors are not atomic data types. Internally in R, factors are stored by numbers (integers), externally the name of the factor is used. We call the corresponding variables *categorical*.

Data types of the hyperparameters that are analyzed in this book can be obtained with the function `getModelConf`. The function `spot` can handle the data types `numeric`, `integer`, and `factor`.

Example: Hyperparameters and Their Types

The following code shows how to get the hyperparameter names and their corresponding types of the *k*-Nearest-Neighbor (KNN) method.

```
library("SPOTMisc")
cfg <- getModelConf(list(model = "kknn"))
cfg$tunepars
```

```
## [1] "k"         "distance"
```

```
cfg$type
```

```
## [1] "integer" "numeric"
```

The method KNN will be described in detail in Sect. 3.2.

Chapter 3
Models

Thomas Bartz-Beielstein and Martin Zaefferer

Abstract This chapter presents a unique overview and a comprehensive explanation of Machine Learning (ML) and Deep Learning (DL) methods. Frequently used ML and DL methods; their hyperparameter configurations; and their features such as types, their sensitivity, and robustness, as well as heuristics for their determination, constraints, and possible interactions are presented. In particular, we cover the following methods: k-Nearest Neighbor (KNN), Elastic Net (EN), Decision Tree (DT), Random Forest (RF), Extreme Gradient Boosting (XGBoost), Support Vector Machine (SVM), and DL. This chapter in itself might serve as a stand-alone handbook already. It contains years of experience in transferring theoretical knowledge into a practical guide.

3.1 Methods and Hyperparameters

In the following, we provide a survey and description of hyperparameters of ML and DL methods. We emphasize that this is not a complete list of their parameters, but covers parameters that are set quite frequently according to the literature.

Since the specific names and meaning of hyperparameters may depend on the actual implementation used, we have chosen a reference implementation for each model. The implementations chosen are all packages from the statistical programming language R. Thus, we provide a description that is consistent with what users experience, so that they can identify the relevant parameters when tuning ML and DL methods in practice. In particular, we cover the methods shown in Table 3.1.

T. Bartz-Beielstein (✉)
Institute for Data Science, Engineering and Analytics, TH Köln, Gummersbach, Germany
e-mail: thomas.bartz-beielstein@th-koeln.de

M. Zaefferer
Bartz & Bartz GmbH and with Institute for Data Science, Engineering, and Analytics, TH Köln, Gummersbach, Germany

Duale Hochschule Baden-Württemberg Ravensburg, Ravensburg, Germany
e-mail: zaefferer@dhbw-ravensburg.de

© The Author(s) 2023
E. Bartz et al. (eds.), *Hyperparameter Tuning for Machine and Deep Learning with R*,
https://doi.org/10.1007/978-981-19-5170-1_3

Table 3.1 Overview: Methods and hyperparameters analyzed in this book

Model, R Package	Hyperparameter	Comment
KNN, kknn	k	Number of neighbors
	p	p norm
EN, glmnet	alpha	Weight term of the loss function
	lambda	Trade-off between model quality and complexity
	thresh	Threshold for model convergence, i.e., convergence of the internal coordinate descent
DT, rpart	minsplit	Minimum number of observations required for a split
	minbucket	Minimum number of observations in an end node (leaf)
	cp	Complexity parameter
	maxdepth	Maximum depth of a leaf in the decision tree
RF, ranger	num.trees	Number of trees that are combined in the overall ensemble model
	mtry	Number of randomly chosen features are considered for each split
	sample.fraction	Number of observations that are randomly drawn for training a specific tree
	replace	Replacement of randomly drawn samples
	respect.-	
	unordered.factors	Handling of splits of categorical variables
XGBoost, xgboost	eta	Learning rate, also called "shrinkage" parameter
	nrounds	Number of boosting steps
	lambda	Regularization of the model
	alpha	Parameter for the L1 regularization of the weights
	subsample	Portion of the observations that is randomly selected in each iteration
	colsample_bytree	Number of features that is chosen for the splits of a tree
	gamma	Number of splits of a tree by assuming a minimal improvement for each split
	maxdepth x	Maximum depth of a leaf in the decision trees
	min_child_weight	Restriction of the number of splits of each tree
SVM, e1071	degree	Degree of the polynomial (parameter of the polynomial kernel function)
	gamma	Parameter of the polynomial, radial basis, and sigmoid kernel functions
	coef0	Parameter of the polynomial and sigmoid kernel functions
	cost	Regularization parameter weighs constraint violations of the model
	epsilon	Regularization parameter defines ribbon around predictions
DL, keras/tensorflow		See Chaps. 8, 9, 10, and 11

This table presents an overview of these methods, their R packages, and associated hyperparameters. After a short, general description of the specific hyperparameter, the following features will be described for every hyperparameter:

Type: Describes the type (e.g., integer) and complexity (e.g., scalar). These data types are described in Sect. 2.7.3. The variable type of the implementation in the R package SPOTMisc, which is used for the experiments in this book, is also listed.

Default: Default value as specified in getModelConf from the R package SPOTMisc.

Sensitivity: Describes how much the model is affected by changes of the parameter. There is a close relationship between sensitivity and tunability as defined by Probst et al. (2019a), because tunability is the potential for improvement of the parameter in the vicinity of a reference value.

Heuristics: Describes ways to find good hyperparameter settings.

Range: Describes feasible values, i.e., lower and upper bounds, constraints, etc.

Transformation: Transformation as specified in getModelConf.

Bounds: Lower and upper bounds as specified in getModelConf.

Constraints: Additional constraints, specific for certain settings or algorithms.

Interactions: Describes interactions between the parameters.

Each description concludes with a brief survey of examples from the literature that gives hints how the method was tuned.

> ! **Attention: Default Hyperparameters**

The default values in this chapter refer to the untransformed values, i.e., the transformations that are also listed in the descriptions were not applied.

3.2 k-Nearest Neighbor

3.2.1 Description

In the field of statistical discrimination KNN classification is an established and successful method. Hechenbichler and Schliep (2004) developed an extended KNN version, where the distances of the nearest neighbors can be taken into account. The KNN model determines for each x the k neighbors with the least distance to x, e.g., based on the Minkowski distance (Eq. (2.1)). For regression, the mean of the neighbors is used (James et al. 2017). For classification, the prediction of the model is the most frequent class observed in the neighborhood. Two relevant hyperparameters (k, p) result from this. Additionally, one categorical hyperparameter could be considered: the choice of evaluation algorithm (e.g., choosing between brute force or KD-Tree) (Friedman et al. 1977). However, this mainly influences computational efficiency, rather than actual performance.

We consider the implementation from the R package kknn[1] (Schliep et al. 2016).

3.2.2 Hyperparameters of k-Nearest Neighbor

KNN Hyperparameter k

The parameter k determines the number of neighbors that are considered by the model. In case of regression, it affects how smooth the predicted function of the model is. Similarly, it influences the smoothness of the decision boundary in case of classification.

Small values of k lead to fairly nonlinear predictors (or decision boundaries), while larger values tend toward more linear shapes (James et al. 2017). The error of the model at any training data sample is zero if k = 1 but this does not allow any conclusions about the generalization error (James et al. 2017). Larger values of k may help to deal with rather noisy data. Moreover, larger values of k increase the runtime of the model.

Type:	integer, scalar.
Default:	7
Sensitivity:	Determining the size of the neighborhood via k is a fairly sensitive decision. James et al. (2017) describe this as a drastic effect. However, this is only true as long as the individual classes are hard to separate (in case of classification). If there is a large margin between classes, the shape of the decision boundary becomes less relevant (see Domingos 2012, Fig. 3). Thus, the sensitivity of the hyperparameter depends on the considered problem and data. Probst et al. (2019a) also identify k as a sensitive (or *tunable*) hyperparameter.
Heuristics:	As mentioned above, the choice of k may depend on properties of the data. Hence, no general rule can be provided. In individual cases, determining the distance between and within classes may help to find an approximate value: k = 1 is better than k > 1, if the distance within classes is larger than the distance between classes (Cover and Hart 1967). Another empirical suggestion from the literature is $k = \sqrt{n}$, where n is the number of data samples (Lall and Sharma 1996; Probst et al. 2019a).
Range:	$k \geq 1$, $k \ll n$. Only integer values are valid.

---

[1] https://cran.r-project.org/package=kknn.

Transformation:	`trans_id`
Bounds:	`lower = 1; upper = 30`
Constraints:	none.
Interactions:	We are not aware of any interactions between the hyperparameters. However, both k and p change the perceived neighborhood of samples and thus the shape of the decision boundaries. Hence, an interaction between these hyperparameters is likely.

KNN Hyperparameter p

The hyperparameter p affects the distance measure that is used to determine the nearest neighbors in KNN. Frequently, this is the Minkowski distance, see Eq. (2.1). Moreover, it has to be considered that other distances could be chosen for non-numerical features of the data set (i.e., Hamming distance for categorical features). The implementation used in the R package kknn transforms categorical variables into numerical variables via dummy-coding, then using the Minkowski distance on the resulting data. Similar to k, p changes the observed neighborhood. While p does not change the number of neighbors, it still affects the choice of neighbors.

Type:	double, scalar.
Default:	`log10(2)`
Sensitivity:	It has to be expected that the model is less sensitive to changes in p than to changes in k, since fairly extreme changes are required to change the neighborhood set of a specific data sample. This explains why many publications do not consider p during tuning, see Table 3.2. However, the detailed investigation of Alfeilat et al. (2019) showed that changes of the distance measure can have a significant effect on the model accuracy. Alfeilat et al. (2019) only tested special cases of the Minkowski distance (Eq. (2.1)): Manhattan distance ($p = 1$), Euclidean distance ($p = 2$) and Chebyshev distance ($p = \infty$). They give no indication whether other values may be of interest as well.
Heuristics:	The choice of distance measure (and hence p) depends on the data, a general recommendation or rule-of-thumb is hard to derive (Alfeilat et al. 2019).
Range:	Often, the interval $1 \leq p \leq 2$ is considered. The lower boundary is $p > 0$. Note: The Minkowski distance is not a metric if $p < 1$ (Alfeilat et al. 2019). Theoretically, a value of $p = \infty$ is possible (resulting into Chebyshev distance), but this is not possible in the kknn implementation.
Transformation:	`trans_10pow`
Bounds:	`lower = -1; upper = 2`
Constraints:	none.

Table 3.2 Survey of examples from the literature, for tuning of KNN

Hyperparameter	Lower bound	Upper bound	Result	Notes
(Schratz et al. 2019), weighted KNN variant, spacial data, 1 data set				
k	10	400	NA	
p	1	100	NA	Integer
(Khan et al. 2020), detection of bugs, 5 data sets				
k	1	17	NA	
p	0,5	5	NA	
(Osman et al. 2017), detection of bugs, 5 data sets				
k	1	5	2 or 5 *	
(Probst et al. 2019a), various applications, 38 data sets				
k	1	30	2 to 30 *	
(Doan et al. 2020), impact damage on reinforced concrete, 1 data set				
k	7	51	9	
p	1	11	3	

*Denotes results that depend on data set (multiple data sets)

Interactions:	We are not aware of any known interactions between hyperparameters. However, both k and p change what is perceived as the neighborhood of samples, and hence the shape of decision boundaries. An interaction between those hyperparameters is likely.

Table 3.2 provides a brief survey of examples from the literature, where KNN was tuned.

3.3 Regularized Regression (Elastic Net)

3.3.1 Description

EN is a regularized regression method (Zou and Hastie 2005). Regularized regression can be employed to fit regression models with a reduced number of model coefficients. Special cases of EN are Lasso and Ridge regression.

Regularization is useful for large k, i.e., when data sets are high dimensional (especially but not exclusively if $k > n$), or when variables in the data sets are heavily correlated with each other (Zou and Hastie 2005). Less complex models (i.e., with fewer coefficients, see also Definition 2.27) help to reduce overfitting. Overfitting means that the model is extremely well adapted to the training data, but generalizes poorly as a result, i.e., predicts poorly for unseen data. The resulting models are also easier to understand for humans, due to their reduced complexity.

During training, non-regularized regression reduces the model error (e.g., via the least squares method), but not the model complexity. EN also considers a penalty term, which grows with the number of coefficients included in the model (i.e., the number of non-zero coefficients).

As a reference implementation, we use the R package `glmnet`[2] (Friedman et al. 2020; Simon et al. 2011).

3.3.2 Hyperparameters of Elastic Net

EN Hyperparameter `alpha`

The parameter `alpha` (α) weighs the two elements of the penalty term of the loss function in the EN model (Friedman et al. 2010):

$$\min_{\beta_0, \beta} \frac{1}{2n} \sum_{i=1}^{n} (y_i - \beta_0 - x_i^{\mathrm{T}} \beta)^2 + \lambda P(\alpha, \beta). \tag{3.1}$$

The penalty term $P(\alpha, \beta)$ is (Friedman et al. 2010)

$$(1 - \alpha) \frac{1}{2} ||\beta||_2^2 + \alpha ||\beta||_1, \tag{3.2}$$

with the vector of p model coefficients $\beta \in \mathbb{R}^p$ and the *intercept* coefficient $\beta_0 \in \mathbb{R}$. The value `alpha` = 0 corresponds to the special case of Ridge regression, `alpha` = 1 corresponds to Lasso regression (Friedman et al. 2010).

The parameter `alpha` allows to find a compromise or trade-off between Lasso and Ridge regression. This can be advantageous, since both variants have different consequences. Ridge regression affects that coefficients of strongly correlated variables match to each other (extreme case: identical variables receive identical coefficients) (Friedman et al. 2010). In contrast, Lasso regression tends to lead to a single coefficient in such a case (the other coefficients being zero) (Friedman et al. 2010).

Type:	double, scalar.
Default:	1
Sensitivity:	Empirical results from Friedman et al. (2010) show that the EN model can be rather sensitive to changes in `alpha`.
Heuristics:	We are not aware of any heuristics to set this parameter. As described by Friedman et al. (2010), `alpha` can be set to a value of close to 1, if a model with few coefficients without risk of degeneration is desired.

---

[2] https://cran.r-project.org/package=glmnet.

Range: `alpha` ∈ [0, 1].
Transformation: `trans_id`
Bounds: `lower = 0; upper = 1`
Constraints: none.
Interactions: `lambda` interacts with `alpha`, see Sect. 3.3.2.

EN Hyperparameter `lambda`

The hyperparameter `lambda` influences the impact of the penalty term $P(\alpha, \beta)$ in Eq. (3.1). Very large `lambda` values lead to many model coefficients (β) being set to zero. Correspondingly, only few model coefficients become zero if `lambda` is small (close to zero). Thus, `lambda` is often treated differently than other hyperparameters: in many cases, several values of `lambda` are of interest, rather than a single value (Simon et al. 2011). There is no singular, optimal solution for `lambda`, as it controls the trade-off between model quality and complexity (number of coefficients that are not zero). Hence, a whole set of `lambda` values will often be suggested to users, who then choose a resulting model that provides a specific trade-off to their liking.

Type: double, scalar.
Default: not implemented, because parameter is not tuned.
Sensitivity: EN is necessarily sensitive to `lambda`, since extreme values lead to completely different models, i.e., all coefficients are zero or none are zero. This is also shown in Fig. 1 by Friedman et al. (2010).
Heuristics: Often, `lambda` gets determined by a type of grid search, where a sequence of decreasing `lambda` is tested (Friedman et al. 2010; Simon et al. 2011). The sequence starts with a sufficiently large value of `lambda`, such that $\beta = 0$. The sequence ends, if the resulting model starts to approximate the unregularized model (Simon et al. 2011).
Range: `lambda` ∈ $(0, \infty)$ (Note: `lambda = 0` is possible, but leads to a simple unregularized model). Using a logarithmic scale seems reasonable, as used in the study by Probst et al. (2019a), to cover a broad spectrum of very small and very large values.
Transformation: not implemented, because parameter is not tuned.
Bounds: not implemented, because parameter is not tuned.
Constraints: none.
Interactions: `lambda` interacts with `alpha`. Both are central for determining the coefficients β (see also Friedman et al. 2010, Fig. 1).

Table 3.3 Survey of examples from the literature, for tuning of EN

Hyperparameter	Lower bound	Upper bound	Result	Notes
(Probst et al. 2019a), various applications, 38 data sets				
alpha	0	1	0,003 to 0,981 *	
lambda	2^{-10}	2^{10}	0,001 to 0,223 *	
(Wong et al. 2019), medical data, 1 data set				
alpha	1	1	1	Not tuned, constant
lambda	**	**	0.001	

*Results depend on data set (multiple data sets)
**The integrated, automatic tuning procedure from `glmnet` was used

EN Hyperparameter `thresh`

The parameter `thresh` is a threshold for model convergence (i.e., convergence of the internal coordinate descent). Model training ends, when the change after an update of the coefficients drops below this value (Friedman et al. 2020). Unlike parameters like `lambda`, `thresh` is not a regularization parameter, hence there is a clear connection between `thresh` and the number of model coefficients.

As a stopping criterion, `thresh` influences the duration of model training (larger values of `thresh` result into faster training), and the quality of the model (larger values of `thresh` may decrease quality).

Type:	double, scalar.
Default:	-7
Sensitivity:	As long as `thresh` is in a reasonable range of values, the model will not be sensitive to changes. Extremely large values can lead to fairly poor models, extremely small values may result into significantly larger training times.
Heuristics:	none are known.
Range:	`thresh` ≈ 0, `thresh` > 0. It seems reasonable to set `thresh` on a log-scale with fairly coarse granularity, since `thresh` has a low sensitivity for the most part. Example: `thresh` $= 10^{-20}, 10^{-18}, \ldots, 10^{-4}$.
Transformation:	`trans_10pow`
Bounds:	`lower = -8; upper = -1`
Interactions:	none are known.

In conclusion, Table 3.3 provides a brief survey of examples from the literature, where EN was tuned.

3.4 Decision Trees

3.4.1 Description

Decision and regression trees are models that divide the data space into individual segments with successive decisions (called splits).

Basically, the procedure of a decision tree is as follows: Starting from a root node (which contains all observations) a first split is carried out. Each split affects a variable (or a feature). This variable is compared with a threshold value. All observations that are less than the threshold are assigned to a new node. All other observations are assigned to another new node. This procedure is then repeated for each node until a termination criterion is reached or until there is only one observation in each end node. End nodes are also called leaves (following the tree analogy).

A detailed description of tree-based models is given by James et al. (2014). An overview of decision tree implementations and algorithms is given by Zharmagambetov et al. (2019). Gomes Mantovani et al. (2018) describe the tuning of hyperparameters of several implementations. As a reference implementation, we refer to the R package `rpart` (Therneau and Atkinson 2019; Therneau et al. 2019).

3.4.2 Hyperparameters of Decision Trees

DT Hyperparameter `minsplit`

If there are fewer than `minsplit` observations in a node of the tree, no further split is carried out at this node. Thus, `minsplit` limits the complexity (number of nodes) of the tree. With large `minsplit` values, fewer splits are made. A suitable choice of `minsplit` can thus avoid overfitting. In addition, the parameter influences the duration of the training of a decision tree (Hastie et al. 2017).

Type:	integer, scalar.
Default:	20
Sensitivity:	Trees can react very sensitively to parameters that influence their complexity. Together with `minbucket`, `cp`, and `maxdepth`, `minsplit` is one of the most important hyperparameters (Gomes Mantovani et al. 2018).
Heuristics:	`minsplit` is set to three times `minbucket` in certain implementations, if this parameter is available (Therneau and Atkinson 2019).
Range:	`minsplit` $\in [1, n]$, where `minsplit` $\ll n$ is recommended, since otherwise trees with extremely few nodes will arise. Only integer values are valid.

Transformation: `trans_id`
Bounds: `lower = 1; upper = 300`
Constraints: `minsplit > minbucket`. This is a soft constraint, i.e., valid
                  models are created even if violated, but `minsplit` would no
                  longer have any effect.
Interactions: The parameters `minsplit`, `minbucket`, `cp`, and `maxdepth`
                  all influence the complexity of the tree. Interactions between these
                  parameters are therefore likely. In addition, `minsplit` has no
                  effect for certain values of `minbucket` (see Constraints). Simi-
                  lar relationships (depending on the data) are also conceivable for
                  the other parameter combinations.

DT Hyperparameter `minbucket`

`minbucket` specifies the minimum number of data points in an end node (leaf) of
the tree. The meaning in practice is similar to that of `minsplit`. With larger values,
`minbucket` also increasingly limits the number of splits and thus the complexity of
the tree. Additional information: `minbucket` is set relative to `minsplit`, i.e., we
are using numerical values for `minbucket` that represent *percentages* relative to
`minsplit`. If `minbucket` = 1.0, then `minbucket` = `minsplit`. `minsplit`
should be greater than or equal `minbucket`.

Type: integer, scalar.
Default: `1/3`
Sensitivity: see `minsplit`.
Heuristics: `minbucket` is set to a third of the values of `minsplit`
                  in the reference implementations, if this parameter is avail-
                  able Therneau et al. (2019).
Range: `minbucket` $\in [1, n]$, where `minbucket` $\ll n$ is recomm-
                  ended, as otherwise trees with extremely few nodes will arise.
                  Only integer values are valid.
Transformation: `trans_id`
Bounds: `lower = 0.1; upper = 0.5`
Constraints: `minsplit > minbucket` (this is a soft constraint, i.e., valid
                  models are created even if violated, but `minsplit` would no
                  longer have any effect).
Interactions: see `minsplit`. Due to the similarity of `minsplit` and
                  `minbucket`, it can make sense to only tune one of the two
                  parameters.

DT Hyperparameter cp

The threshold complexity cp controls the complexity of the model in that split decisions are linked to a minimal improvement. This means that if a split does not improve the tree-based model by at least the factor cp, this split will not be carried out. With larger values, cp increasingly limits the number of splits and thus the complexity of the tree.

Therneau and Atkinson (2019) describe the cp parameter as follows:

The complexity parameter cp is, like minsplit, an advisory parameter, but is considerably more useful. It is specified according to the formula

$$R_{cp}(T) \equiv R(T) + cp \times |T| \times R(T_1), \qquad (3.3)$$

where T_1 is the tree with no splits, $|T|$ is the number of splits for a tree, and R is the risk. This scaled version is much more user-friendly than the original CART formula since it is unit less. A value of cp = 1 will always result in a tree with no splits. For regression models, the scaled cp has a very direct interpretation: if any split does not increase the overall R^2 of the model by at least cp (where R^2 is the usual linear models definition) then that split is decreed to be, a priori, not worth pursuing. The program does not split said branch any further and saves considerable computational effort. The default value of 0.01 has been reasonably successful at "pre-pruning" trees so that the cross-validation step only needs to remove one or two layers, but it sometimes over prunes, particularly for large data sets.

Type:	double, scalar.
Default:	-2
Sensitivity:	see minsplit.
Heuristics:	none known.
Range:	paramcp ∈ [0, 1[.
Transformation:	trans_10pow
Bounds:	lower = -10; upper = 0
Constraints:	none.
Interactions:	see minsplit. Since cp expresses a relative factor for the improvement of the model, an interaction with the corresponding quality measure is also possible (split parameter).

DT Hyperparameter maxdepth

The parameter maxdepth limits the maximum depth of a leaf in the decision tree. The depth of a leaf is the number of nodes that lie on the path between the root and the leaf. The root node itself is not counted (Therneau and Atkinson 2019).

The meaning in practice is similar to that of minsplit. Both minsplit and maxdepth can be used to limit the complexity of the tree. However, smaller values of maxdepth lead to a lower complexity of the tree. With minsplit it is the other way round (larger values lead to less complexity).

Table 3.4 DT: survey of examples from the literature. Tree-based tuning example configurations

Hyperparameter	Lower bound	Upper bound	Result
(Probst et al. 2019a), various applications, 38 data sets			
minsplit	1	60	6.7 to 49.15 *
maxdepth	1	30	9 to 28 *
cp	0	1	0 to 0.528 *
minbucket	1	60	1 to 44.1 *
(Wong et al. 2019), medical data, 1 data set			
cp	10^{-6}	10^{-1}	10^{-2}
(Khan et al. 2020), software bug detection, 5 data sets			
minbucket	1	50	NA
(Gomes Mantovani et al. 2018), various data sets, 94 data sets			
minsplit	1	50	
minbucket	1	50	
cp	0.0001	0.1	
maxdepth	1	50	
This study, see Chap. 8, Census-Income (KDD) Data Set (CID)			
minsplit	1	300	16 (not relevant)
minbucket	0.1	0.5	0.17 (not relevant)
cp	10^{-10}	1	10^{-3} (most relevant hyperparameter)
maxdepth	1	30	>10

*Denotes that results depend on data sets

Type:	integer, scalar.
Default:	30
Sensitivity:	see minsplit.
Heuristics:	none known.
Range:	maxdepth $\in [0, n]$. Only integer values are valid.
Transformation:	trans_id.
Bounds:	lower = 1; upper = 30.
Constraints:	none.
Interactions:	see minsplit.

Table 3.4 shows examples from the literature.

3.5 Random Forest

3.5.1 Description

The model quality of decision trees can often be improved with ensemble methods. Here, many individual models (i.e., many individual trees) are merged into one overall model (the ensemble). Popular examples are RF and XGBoost methods. This section discusses RF methods, XGBoost methods will be discussed in Sect. 3.6. The RF method creates many decision trees at the same time, and their prediction is then usually made using the mean (in case of regression) or by majority vote (in case of classification).

The variant of RF described by Breiman (2001) uses two important steps to reduce generalization error: first, when creating individual trees, only a random subset of the features is considered for each split. Second, each tree is given a randomly drawn subset of the observations to train. Typically, the approach of bootstrap aggregating or bagging (James et al. 2017) is used. A comprehensive discussion of random forest models is provided by Louppe (2015), who also presents a detailed discussion of hyperparameters. Theoretical results on hyperparameters of RF models are summarized by Scornet (2017). Often, tuning of RF also takes into account parameters for the decision trees themselves as described in Sect. 3.4. Our reference implementation studied in this report is from the R package `ranger`[3] (Wright 2020; Wright and Ziegler 2017).

3.5.2 Hyperparameters of Random Forests

RF Hyperparameter `num.trees`

`num.trees` determines the number of trees that are combined in the overall ensemble model. In practice, this influences the quality of the method (more trees improve the quality) and the runtime of the model (more trees lead to longer runtimes for training and prediction).

Type:	integer, scalar.
Default:	`log(500,2)`.
Sensitivity:	According to Breiman (2001), the generalization error of the model converges with increasing number of trees toward a lower bound. This means that the model will become less sensitive to changes of `num.trees` with increasing values of `num.trees`. This is also shown in the benchmarks of Louppe (2015). Only with relatively small values (`num.trees` < 50) the model is

---

[3] https://cran.r-project.org/package=ranger.

	rather sensitive to changes in that parameter. The empirical results of Probst et al. (2019a) also show that the tunability of num.trees is estimated to be rather low.
Heuristics:	There are theoretical results about the convergence of the model in relation to num.trees (Breiman 2001; Scornet 2017). This however does not result in a clear heuristic approach to setting this parameter. One common recommendation is to choose num.trees sufficiently high (Probst et al. 2019c) (since more trees are usually better), while making sure that the runtime of the model does not become too large.
Range:	num.trees $\in [1, \infty)$. Several hundred or thousands of trees are commonly used, see also Table 3.5.
Transformation:	trans_2pow_round.
Bounds:	lower = 0; upper = 11.
Constraints:	none.
Interactions:	none are known.

RF Hyperparameter mtry

The hyperparameter mtry determines how many randomly chosen features are considered for each split. Thus, it controls an important aspect, the randomization of individual trees. Values of mtry $\ll n$ imply that differences between trees will be larger (more randomness). This increases the potential error of individual trees, but the overall ensemble benefits (Breiman 2001; Louppe 2015). As a useful side effect mtry $\ll n$ may also reduce the runtime considerably (Louppe 2015). Nevertheless, findings about this parameter largely depend on heuristics and empirical results. According to Scornet (2017), no theoretical results about the randomization of split features are available.

Type:	integer, scalar.
Default:	floor(sqrt(nFeatures)).
Sensitivity:	According to Breiman (2001), RF is relatively insensitive to changes of mtry: "But the procedure is not overly sensitive to the value of F. The average absolute difference between the error rate using F=1 and the higher value of F is less than 1." (Breiman 2001) (here: F corresponds to mtry).
	This seems to be at odds with the benchmarks by Louppe (2015), which determine that mtry may indeed have a considerable impact, especially for low values of mtry. The investigation of tunability by Probst et al. (2019a) also identifies mtry as an important (i.e., tunable) parameter. This is not necessarily a contradiction to Breiman's observation, since Probst et al. (2019a) determine RF as the least tunable model in their experimental

Heuristics:

investigation. So while `mtry` might have some impact (compared to other parameters), it may be less sensitive when compared in relation to hyperparameters of other models.

Breiman (2001) propose the following heuristic:

$$mtry = \text{floor}(\log_2(n) + 1).$$

Should categorical features be present, Breiman suggests doubling or tripling that value. No theoretical motivation is given. Another frequent suggestion is $mtry = \sqrt{n}$ (or $mtry = $ floor $(\sqrt{n})$). While these are used in various implementations of RF, there is no clear theoretical motivation given. For $n < 20$ both heuristics provide very similar values.

Some implementations distinguish between classification ($mtry = \sqrt{n}$) and regression ($mtry = n/3$). Empirical results with these heuristics are described by Probst et al. (2019c).

Range: $mtry \in [1, n]$.
Transformation: `trans_id`.
Bounds: `lower = 1; upper = nFeatures`.
Constraints: none.
Interactions: none are known.

RF Hyperparameter `sample.fraction`

The parameter `sample.fraction` determines how many observations are randomly drawn to train one specific tree.

Probst et al. (2019c) write that `sample.fraction` has a similar effect as `mtry`. That means, it influences the properties of the trees: With small `sample.fraction` (corresponding to small `mtry`) individual trees are weaker (in terms of predictive quality), yet the diversity of trees is increased. This improves the ensemble model quality. Smaller values of `sample.fraction` reduce the runtime (Probst et al. 2019c) (if all other parameters are equal).

Type: double, scalar.
Default: 1.
Sensitivity: `sample.fraction` can have a relevant impact on model quality. Scornet reports: "However, according to empirical results, there is no justification for default values in random forests for sub-sampling or tree depth, since optimizing either leads to better performance."
Heuristics: none known.
Range: `sample.fraction` $\in (0, 1]$.
Transformation: `trans_id`.

Bounds:	`lower = 0.1; upper = 1.`
Constraints:	none.
Interactions:	Potentially, `sample.fraction` interacts with parameters that influence training individual trees DT (e.g., `maxdepth`, `minsplit`, `cp`). Scornet: "According to the theoretical analysis of median forests, we know that there is no need to optimize both the `subsample` size and the tree depth: optimizing only one of these two parameters leads to the same performance as optimizing both of them" (Scornet 2017). However, this theoretical observation is only valid for the respective *median trees* and not necessarily for the classical RF model we consider.

RF Hyperparameter `replace`

The parameter `replace` specifies, whether randomly drawn samples are replaced, i.e., whether individual samples can be drawn multiple times for training of a tree (`replace = TRUE`) or not (`replace = FALSE`). If `replace = TRUE`, the probability that two trees receive the same data sample is reduced. This may further decorrelate trees and improve quality.

Type:	logical, scalar.
Default:	2 (TRUE).
Sensitivity:	The sensitivity of `replace` is often rather small. Yet, the survey of Probst et al. (2019c) notes a potentially detrimental bias for `replace = TRUE`, if categorical variables with a variable number of levels are present.
Heuristics:	Due to the aforementioned bias, the choice could be made depending on the variance of the cardinality in the data features. However, a quantifiable recommendation is not available.
Range:	`replace ∈ {TRUE, FALSE}`.
Transformation:	`trans_id`.
Bounds:	`lower = 1 (FALSE); upper = 2 (TRUE)`.
Constraints:	none.
Interactions:	One obvious interaction occurs with `sample.fraction`. Both parameters control the random choice of training data for each tree. The setting (`replace = TRUE ∧ sample.fraction = 1`) as well as the setting (`replace = FALSE ∧ sample.fraction < 1`) implies that individual trees will not see the whole data set.

RF Hyperparameter `respect.unordered.factors`

This parameter decides how splits of categorical variables are handled. There are basically three options: `ignore`, `order`, or `partition`, which will briefly be explained in the following. A detailed discussion is given by Wright and König (2019). A standard that is also used by Breiman (2001) is `respect.unordered.factors = partition`. In that case, all potential splits of a nominal, categorical variable are considered. This leads to a good model, but the large number of considered splits can lead to an unfavorable runtime.

A naive alternative is `respect.unordered.factors = ignore`. Here, the categorical nature of a variable will be ignored. Instead, it is assumed that the variable is ordinal, and splits are chosen just as with numerical variables. This reduces runtime but can decrease model quality.

A better choice should be `respect.unordered.factors = order`. Here, each categorical variable first is sorted, depending on the frequency of each level in the first of two classes (in case of classification) or depending on the average dependent variable value (regression). After this sorting, the variable is considered to be numerical. This allows for a runtime similar to that with `respect.unordered.factors = ignore` but with potentially better model quality. This may not be feasible for classification with more than two classes, due to lack of a clear sorting criterion (Wright and König 2019; Wright 2020).

In specific cases, `respect.unordered.factors = ignore` may work well in practice. This could be the case, when the variable is actually nominal (unknown to the analyst).

Type:	character, scalar.
Default:	1 (`ignore`).
Sensitivity:	unknown.
Heuristics:	none.
Range:	`respect.unordered.factors` ∈ {`ignore`, `order`, `partition`}. The parameter `respect.unordered.factors` can also be understood as a binary value. Then TRUE corresponds to `order` and FALSE to `ignore` (Wright 2020).
Transformation:	`trans_id`.
Bounds:	lower = 1 (`ignore`); upper = 2 (`order`).
Constraints:	none.
Interactions:	none are known.

In conclusion, Table 3.5 provides a brief survey of examples from the literature, where RF was tuned.

Table 3.5 RF: survey of examples from the literature for tuning of random forest

Hyperparameter	Lower bound	Upper bound	Result	Notes
(Probst et al. 2019a), various applications, 38 data sets				
`num.trees`	1	2000	187,85 to 1908,25 *	
`replace`			FALSE	Binary
`sample.fraction`	0.1	1	0,257 to 0,974 *	
`mtry`	0	1	0,035 to 0,954 *	Transformed: `mtry` $\times m$
`respect.unordered.factors`			FALSE oder TRUE	binary
min.node.size	0	1	0,007 to 0,513 *	Transformed: $n^{\text{min.node.size}}$
(Schratz et al. 2019), spatial data, 1 data set				
`num.trees`	10	10000	NA	
`mtry`	1	11	NA	
(Wong et al. 2019), medical data, 1 data set				
`num.trees`	10	2000	1000	
`mtry`	10	200	50	

*Results depend on data set (multiple data sets)

3.6 Gradient Boosting (xgboost)

3.6.1 Description

Boosting is an ensemble process. In contrast to random forests, see Sect. 3.5, the individual models (here: decision trees) are not created and evaluated at the same time, but rather sequentially. The basic idea is that each subsequent model tries to compensate for the weaknesses of the previous models.

For this purpose, a model is created repeatedly. The model is trained with weighted data. At the beginning these weights are identically distributed. Data that are poorly predicted or recognized by the model are given larger weights in the next step and thus have a greater influence on the next model. All models generated in this way are combined as a linear combination to form an overall model (Freund and Schapire 1997; Drucker and Cortes 1995).

An intuitive description of this approach is *slow learning*, as the attempt is not made to understand the entire database in a single step, but to improve the understanding step by step (James et al. 2014). Gradient Boosting (GB) is a variant of this approach, with one crucial difference: instead of changing the weighing of the data, models are created sequentially that follow the gradient of a loss function. In the case of regression, the models learn with residuals of the sum of all previous models. Each individual model tries to reduce the weaknesses (here: residuals) of the ensemble (Friedman 2001).

In the following, we consider the hyperparameters of one version of GB: XGBoost (Chen and Guestrin 2016). In principle, any models can be connected in ensembles via boosting. We apply XGBoost to decision trees. As a reference implementation, we refer to the R package xgboost (Chen et al. 2020). Brownlee (2018) describes some empirical hyperparameter values for tuning XGBoost.

3.6.2 Hyperparameters of Gradient Boosting

XGBoost Hyperparameter `nrounds`

The parameter `nrounds` specifies the number of boosting steps. Since a tree is created in each individual boosting step, `nrounds` also controls the number of trees that are integrated into the ensemble as a whole. Its practical meaning can be described as follows: larger values of `nrounds` mean a more complex and possibly more precise model, but also cause a longer running time. The practical meaning is therefore very similar to that of num.trees in random forests. In contrast to num.trees, overfitting is a risk with very large values, depending on other parameters such as `eta`, `lambda`, `alpha`. For example, the empirical results of Friedman (2001) show that with a low `eta`, even a high value of `nrounds` does not lead to overfitting.

Type:	integer, scalar.
Default:	0.
Sensitivity/robustness	Similar to the random forests parameter num.trees, `nrounds` also has a higher sensitivity, especially with low values (Friedman 2001).
Heuristics:	Heuristics cannot be derived from the literature. Often values of several hundred to several thousand trees are set as the upper limit (Brownlee 2018).
Range:	$\in [1, \infty[$. Only integer values are valid.
Transformation:	`trans_2pow_round`.
Bounds:	`lower = 0; upper = 11`.
Constraints:	none.
Interactions:	There is a connection between the hyperparameters beta, rounds, and `subsample`.

XGBoost Hyperparameter `eta`

The parameter `eta` is a learning rate and is also called "shrinkage" parameter. It controls the lowering of the weights in each boosting step (Chen and Guestrin 2016; Friedman 2002). It has the following practical meaning: lowering the weights helps to reduce the influence of individual trees on the ensemble. This can also avoid overfitting (Chen and Guestrin 2016).

Type: double, scalar.

Default: `log2(0.3)`.

Sensitivity: Empirical results show that XGBoost is more sensitive to `eta` when `eta` is large (Friedman 2001). Generally speaking, smaller values are better. In an empirical study, Probst et al. (2018) describe `eta` as a parameter with comparatively high tunability.

Heuristics: A heuristic is difficult to formulate due to the dependence on other parameters and the data situation, but Hastie et al. (2017) recommend

> … the best strategy appears to be to set `eta` to be very small (eta < 0.1) and then choose `nrounds` by early stopping.

This may lead to correspondingly longer runtimes due to large `nrounds`. Brownlee (2018) mentions a heuristic, which describes a search range depending on `nrounds`.

Range: $eta \in [0, 1]$. Using a logarithmic scale seems reasonable, e.g., $2^{-10}, \ldots, 2^0$), as used in the studies by Probst et al. (2018) or Sigrist (2020), because values close to zero often show good results.

Transformation: `trans_2pow`.

Bounds: `lower = -10; upper = 0` .

Constraints: none.

Interactions: There is a connection between `eta` and `nrounds`: If one of the two parameters increases, the other should be decreased if the error remains the same (Friedman 2001; Probst et al. 2019a). This is also demonstrated by Hastie et al. (2017):

> Smaller values of `eta` lead to larger values of `nrounds` for the same training risk, so that there is a trade-off between them.

In addition, Hastie et al. (2017) also point to correlations with the `subsample` parameter: In an empirical study, `subsample` = 1 and `eta` = 1 show significantly worse results than `subsample` = 0.5 and `eta` = 0.1. If `subsample` = 0.5 and `eta` = 1, the results are even worse than for `eta` = 1 and `subsample` = 1. In the best case (`subsample` = 0.5 and `eta` = 0.1), however, larger values of `nrounds` are required to achieve optimal results.

XGBoost Hyperparameter `lambda`

The parameter `lambda` is used for the regularization of the model. This parameter influences the complexity of the model (Chen and Guestrin 2016; Chen et al. 2020) (similar to the parameter of the same name in EN). Its practical significance can be described as follows: as a regularization parameter, `lambda` helps to prevent overfitting (Chen and Guestrin 2016). With larger values, smoother or simpler models are to be expected.

Type:	double, scalar.
Default:	0.
Sensitivity:	not known.
Heuristics:	none known.
Range:	`lambda` $\in [0, \infty[$. A logarithmic scale seems to be useful, e.g., $2^{-10}, \ldots, 2^{10}$, as used in the study by Probst et al. (2019a) to cover a wide range of very small and very large values.
Transformation:	`trans_2pow`.
Bounds:	`lower = -10; upper = 10`.
Constraints:	none.
Interactions:	Because both `lambda` and `alpha` control the regularization of the model, an interaction is likely.

XGBoost Hyperparameter `alpha`

The authors of the R package `xgboost`, Chen and Guestrin (2016), did not mention this parameter. The documentation of the reference implementation does not provide any detailed information on `alpha` either. Due to the description as a parameter for the l_1 regularization of the weights (Chen et al. 2020), a highly similar use as for the parameter of the same name in elastic net is to be assumed. Its practical meaning can be described as follows: similar to `lambda`, `alpha` also functions as a regularization parameter.

Type:	double, scalar.
Default:	-10.
Sensitivity:	unknown.
Heuristics:	No heuristics are known.
Range:	`alpha` $\in [0, \infty[$. A logarithmic scale seems to be useful, e.g., $2^{-10}, \ldots, 2^{10}$, as used in the study by Probst et al. (2019a) to cover a wide range of very small and very large values.
Transformation:	`trans_2pow`.
Bounds:	`lower = -10; upper = 10`.
Constraints:	none.
Interactions:	Since both `lambda` and `alpha` control the regularization of the model, an interaction is likely.

XGBoost Hyperparameter `subsample`

In each boosting step, the new tree to be created is usually only trained on a subset of the entire data set, similar to random forest (Friedman 2002). The `subsample` parameter specifies the portion of the data approach that is randomly selected in each iteration. Its practical significance can be described as follows: an obvious effect of small `subsample` values is a shorter running time for the training of individual trees, which is proportional to the `subsample` (Hastie et al. 2017).

Type:	double, scalar.
Default:	1.
Sensitivity:	The study by Friedman (2002) shows a high sensitivity for very small or large values of `subsample`. In a relatively large range of values from `subsample` (around 0.3 to 0.6), however, hardly any differences in model quality are observed.
Determination heuristics:	Hastie et al. (2017) suggest `subsample` = 0.5 as a good starting value, but point out that this value can be reduced if `nrounds` increases. With many trees (nround is large) it is sufficient if each individual tree sees a smaller part of the data, since the unseen data is more likely to be taken into account in other trees.
Range:	`subsample` ∈]0, 1]. Based on the empirical results Friedman (2002); Hastie et al. (2017), a logarithmic scale is not recommended.
Transformation:	`trans_id`.
Bounds:	`lower = 0.1; upper = 1`.
Constraints:	none.
Interactions:	There is a connection between the `eta`, `nrounds`, and `subsample`.

XGBoost Hyperparameter `colsample_bytree`

The parameter `colsample_bytree` has similarities to the mtry parameter in random forests. Here, too, a random number of features is chosen for the splits of a tree. In XGBoost, however, this choice is made only once for each tree that is created, instead for each split (xgboost developers 2020). Here `colsample_bytree` is a relative factor. The number of selected features is therefore `colsample_bytree` $\times n$. Its practical meaning is similar to mtry: `colsample_bytree` enables the trees of the ensemble to have a greater diversity. The runtime is also reduced, since a smaller number of splits have to be checked each time (if `colsample_bytree` < 1).

Type:	double, scalar.
Default:	1.
Sensitivity:	The empirical study by Probst et al. (2019a) shows that the model is particularly sensitive to changes for `colsample_bytree` values close to 1. However, this sensitivity decreases in the vicinity of more suitable values.
Heuristics:	none known.
Range:	`colsample_bytree` ∈]0, 1]. Brownlee (2018) mentions search ranges such as `colsample_bytree` = 0.4, 0.6, 0.8, 1, but mostly works with `colsample_bytree` = 0.1, 0.2, ..., 1.
Transformation:	`trans_id`.
Bounds:	`lower = 1/nFeatures; upper = 1.`
Constraints:	none.
Interactions:	none known.

XGBoost Hyperparameter `gamma`

This parameter of a single decision tree is very similar to the parameter `cp`: Like `cp`, `gamma` controls the number of splits of a tree by assuming a minimal improvement for each split. According to the documentation (Chen et al. 2020):

> Minimum loss reduction required to make a further partition on a leaf node of the tree. The larger, the more conservative the algorithm will be.

The main difference between `cp` and `gamma` is the definition of `cp` as a relative factor, while `gamma` is defined as an absolute value. This also means that the ranges differ.

Default:	-10.
Range:	`gamma` ∈ [0, ∞[. A logarithmic scale seems to make sense, e.g., $2^{-10}, \ldots, 2^{10}$, as, e.g., in the study by Thomas et al. (2018) to cover a wide range of very small and very large values.
Transformation:	`trans_2pow`.
Bounds:	`lower = -10; upper = 10.`

XGBoost Hyperparameter `maxdepth`

This parameter of a single decision tree is already known as `maxdepth`.

Default:	6.

Sensitivity/heuristics: Hastie et al. (2017) state:

> Although in many applications $J = 2$ will be insufficient, it is unlikely that $J > 10$ will be required. Experience so far indicates that $4 \leq J \leq 8$ works well in the context of boosting, with results being fairly insensitive to particular choices in this range.[4]

Transformation: `trans_id`.
Bounds: `lower = 1; upper = 15`.

XGBoost Hyperparameter `min_child_weight`

Like `gamma` and `maxdepth`, `min_child_weight` restricts the number of splits of each tree. In the case of `min_child_weight`, this restriction is determined using the Hessian matrix of the loss function (summed over all observations in each new terminal node) (Chen et al. 2020; Sigrist 2020). In experiments by Sigrist (2020), this parameter turns out to be comparatively difficult to tune: the results show that tuning with `min_child_weight` gives worse results than tuning with a similar parameter (limitation of the number of samples per sheet) (Sigrist 2020).

Type: double, scalar.
Default: `0`.
Sensitivity: unknown.
Heuristics: none known.
Range: `min_child_weight` $\in [0, \infty[$. A logarithmic scale seems to make sense, e.g., $2^{-10}, \ldots, 2^{10}$, as used in the study by Probst et al. (2019a) to cover a wide range of very small and very large values.
Transformation: `trans_2pow`.
Bounds: `lower = 0; upper = 7`.
Constraints: none.
Interactions: Interactions with parameters such as `gamma` and `maxdepth` are probable, since all three parameters influence the complexity of the individual trees in the ensemble.

Table 3.6 shows XGBoost example parameter settings from the literature.

---

[4] J is the number of nodes in a tree that is strongly influenced by `maxdepth`.

Table 3.6 Survey: examples from literature about XGBoost tuning

Hyperparameter	Lower bound	Upper bound	Result
(Probst et al. 2019a), several applications, 38 data sets			
nrounds	1	5000	920.7 to 4847.15 *
eta	2^{-10}	2^0	0.002 to 0.445 *
subsample	0.1	1	0.545 to 0.964 *
maxdepth x	1	15	2.6 to 14 *
min_child_weight	2^0	2^7	1.061 to 7.502 *
colsample_bytree	0	1	0.334 to 0.922 *
lambda	2^{-10}	2^{10}	0.004 to 29.755 *
alpha	2^{-10}	2^{10}	0.002 to 6.105 *
(Thomas et al. 2018), several applications, 16 data sets			
eta	0.01	0.2	
gamma	2^{-7}	2^6	
subsample	0.5	1	
maxdepth x	3	20	
colsample_bytree	0.5	1	
lambda	2^{-10}	2^{10}	
alpha	2^{-10}	2^{10}	
(Wang 2019), Risk Classification, 1 data set			
eta	0.005	0.2	
subsample	0.8	1	
maxdepth x	5	30	
min_child_weight	0	10	
colsample_bytree	0.8	1	
gamma	0	0.02	
(Zhou et al. 2020), Tunnel construction, 1 data set			
nrounds	1	150	103
eta	0.00001	1	0.152
maxdepth x	1	15	15
lambda	1	15	13
alpha	1	15	1
This study, see Sect. 9.1, CID			
nrounds	0	32	256
eta	2^{-10}	0	0.125

*Denotes that results depend on the data (several data sets)

3.7 Support Vector Machines

3.7.1 Description

The SVM is a kernel-based method.

Definition 3.1 (*Kernel*) A kernel is a real-valued, symmetrical function $k(x, x')$ (usually positive definite), which often expresses some form of similarity between two observations x, x'.

The usefulness of kernels can be explained by the Kernel-Trick. The Kernel-Trick describes the ability of kernels to transfer data into a higher dimensional feature space. This allows classification with linear decision boundaries (hyperplanes) even in cases where the data in the original feature space are not linearly separable (Schölkopf and Smola 2001).

As reference implementation, we use the R package e1071[5] (Meyer et al. 2020), which is based on libsvm (Chang and Lin 2011).

3.7.2 Hyperparameters of the SVM

SVM Hyperparameter `kernel`

The parameter `kernel` is central for the SVM model. It describes the choice of the function $k(x, x')$. In practice, $k(x, x')$ can often be understood to be a measure of similarity. That is, the kernel function describes how similar two observations are to each other, depending on their feature values.

Type:	character, scalar.
Default:	1 (`radial`).
Sensitivity:	The empirical investigation of Probst et al. (2019a) shows "In svm the biggest gain in performance can be achieved by tuning the kernel, `gamma` or `degree`, while the cost parameter does not seem to be very tunable." This does not necessarily mean that `cost` should not be tuned, as the tunability investigated by Probst et al. (2019a) always considers a reference value (e.g., the default).
Heuristics:	Informally, it is often recommended to use `kernel = radial basis`. This also matches well to results and observations from the literature (Probst et al. 2019a; Guenther and Schonlau 2016). With very large numbers of observations and/or features Hsu et al. (2016) suggest to use `kernel = linear`. These are infallible

---

[5] https://cran.r-project.org/package=e1071.

	rules, other kernels may perform better depending on the data set. This stresses the necessity of using hyperparameter tuning to choose kernels.
Range:	• linear: $k(x, x') = x^T x'$.
	• polynomial: $k(x, x') = (\text{gamma } x^T x' + \text{coef0})^{\text{degree}}$.
	• radial basis: $k(x, x') = \exp(-\text{gamma } \|x - x'\|^2)$.
	• sigmoid: $k(x, x') = \tanh(\text{gamma } x^T x' + \text{coef0})$.
Transformation:	`trans_id`.
Bounds:	`lower = 1 (radial); upper = 2 (sigmoid)`.
Constraints:	none.
Interactions:	The kernel functions themselves have parameters (`degree`, `gamma`, and `coef0`), whose values only matter if the respective function is chosen.

SVM Hyperparameter `degree`

The parameter `degree` influences the kernel function if a polynomial kernel was selected:

• polynomial: $k(x, x') = (\text{gamma } x^T x' + \text{coef0})^{\text{degree}}$.

Integer values of `degree` determine the degree of the polynomial. Non-integer values are possible, even though not leading to a polynomial in the classical sense. If `degree` has a value close to one, the polynomial kernel approximates the linear kernel. Else, the kernel becomes correspondingly nonlinear.

Type:	double, scalar.
Default:	not implemented, because parameter is not tuned.
Sensitivity:	The empirical investigation of Probst et al. (2019a) shows "In svm the biggest gain in performance can be achieved by tuning the kernel, `gamma` or `degree`, while the cost parameter does not seem to be very tunable."
Heuristics:	none are known.
Range:	$\text{degree} \in (0, \infty)$.
Transformation:	not implemented, because parameter is not tuned.
Bounds:	not implemented, because parameter is not tuned.
Constraints:	none.
Interactions:	The parameter only has an impact if `kernel = polynomial`.

SVM Hyperparameter gamma

The parameter gamma influences three kernel functions:

- polynomial: $k(x, x') = (\text{gamma } x^T x' + \text{coef0})^{\text{degree}}$.
- radial basis: $k(x, x') = \exp(-\text{gamma } ||x - x'||^2)$.
- sigmoid: $k(x, x') = \tanh(\text{gamma } x^T x' + \text{coef0})$.

In case of polynomial and sigmoid, gamma acts as a multiplier for the scalar product of two feature vectors. For radial basis, gamma acts as a multiplier for the distance of two feature vectors.

In practice, gamma scales how far the impact of a single data sample reaches in terms of influencing the model. With small gamma values, an individual observation may potentially influence the prediction in a larger vicinity, since with increasing distance between x and x', their similarity will decrease more slowly (esp. with kernel = radial basis).

Type:	double, scalar.
Default:	log(1/nFeatures,2).
Sensitivity:	The empirical investigation of van Rijn and Hutter (2018) shows that gamma is rather sensitive.
Heuristics:	The reference implementation uses a simple heuristic, to determine gamma: gamma $= 1/n$ (Meyer et al. 2020). Another implementation (the sigest function in kernlab[6]) first scales all input data, so that each feature has zero mean and unit variance. Afterward, a good interval for gamma is determined, by using the 10 and 90% quantile of the distances between the scaled data samples. By default, 50% randomly chosen samples from the input data are used.
Range:	gamma $\in [0, \infty)$. Using a logarithmic scale seems reasonable (e.g., $2^{-10}, \ldots, 2^{10}$ as used by Probst et al. 2019a), to cover a broad spectrum of very small and very large values.
Transformation:	trans_2pow.
Bounds:	lower = -10; upper = 10.
Constraints:	none.
Interactions:	This parameter has no effect when kernel = linear. In addition, empirical results show a clear interaction with cost (van Rijn and Hutter 2018).

---

[6] https://cran.r-project.org/package=kernlab.

SVM Hyperparameter `coef0`

The parameter `coef0` influences two kernel functions:

- polynomial: $k(x, x') = (\text{gamma } x^T x' + \text{coef0})^{\text{degree}}$.
- sigmoid: $k(x, x') = \tanh(\text{gamma } x^T x' + \text{coef0})$.

In both cases, `coef0` is added to the scalar product of two feature vectors.

Type:	double, scalar.
Default:	0.
Sensitivity:	Empirical results of Zhou et al. (2011) show that `coef0` has a strong impact in case of the polynomial kernel (but only for `degree` = 2).
Heuristics:	Guenther and Schonlau (2016) suggest to leave this parameter at `coef0` = 0.
Range:	`coef0` $\in \mathbb{R}$.
Transformation:	`trans_id`.
Bounds:	`lower` = -1; `upper` = 1
Constraints:	none.
Interactions:	This parameter is only active if `kernel` = polynomial or `kernel` = sigmoid.

SVM Hyperparameter `cost`

The parameter `cost` (often written as C) is a constant that weighs constraint violations of the model. C is a typical regularization parameter, which controls the complexity of the model (Cherkassky and Ma 2004), and may help to avoid overfitting or dealing with noisy data.[7]

Type:	double, scalar.
Default:	0.
Sensitivity:	The empirical results of van Rijn and Hutter (2018) show that `cost` has a strong impact on the model, while the investigation of Probst et al. (2019a) determines only a minor tunability. This disagreement may be explained, since `cost` may have a huge impact in extreme cases, yet good parameter values are found close to the default values.
Heuristics:	Cherkassky and Ma (2004) suggest the following: `cost` = $\max(\lvert \bar{y} + 3\sigma_y \rvert, \lvert \bar{y} - 3\sigma_y \rvert)$. Here, $\bar{y}$ is the mean of the observed y values in the training data, and σ_y is the standard deviation.

---

[7] Here, complexity does not mean the number of model coefficients (as in linear models) or splits (decision trees), but the potential to generate more active/rugged functions. In that context, C influences the number of support vectors in the model. A high model complexity (many support vectors) can create functions with many peaks. This may lead to overfitting.

They justify this heuristic, by pointing out a connection between `cost` and the predicted y: as a constraint, `cost` limits the output values of the SVM model (regression) and should hence be set in a similar order of magnitude as the observed y (Cherkassky and Ma 2004).

Range: `cost` $\in [0, \infty)$. Using a logarithmic scale seems reasonable (e.g., $2^{-10}, \ldots, 2^{10}$ as used by Probst et al. (2019a)), to cover a broad spectrum of very small and very large values.

Transformation: `trans_2pow`.

Bounds: `lower = -10; upper = 10`

Constraints: none.

Interactions: Empirical results show a clear interaction with `gamma` (van Rijn and Hutter 2018).

SVM Hyperparameter `epsilon`

The parameter `epsilon` defines a corridor or "ribbon" around predictions. Residuals within that ribbon are tolerated by the model, i.e., are not penalized (Schölkopf and Smola 2001). The parameter is only used for regression with SVM, not for classification. In the experiments in Sect. 12.1, `epsilon` is only considered when SVM is used for regression.

Similar to `cost`, `epsilon` is a regularization parameter. With larger values, `epsilon` allows for larger errors/residuals. This reduces the number of support vectors (and incidentally, also the runtime). The model becomes more smooth (cf. Schölkopf and Smola 2001, Fig. 9.4). This can be useful, e.g., to deal with noisy data and avoid overfitting. However, the model quality may be decreased.

Type: double, scalar.

Default: `-1`.

Sensitivity: As described above, `epsilon` has a significant impact on the model.

Heuristics: For SVM regression, Cherkassky and Ma (2004) suggest based on simplified assumptions and empirical results: `epsilon` $= 3\sigma\sqrt{\frac{\ln(n)}{n}}$. Here, σ^2 is the noise variance, which has to be estimated from the data, see, e.g., Eqs. (22), (23), and (24) in Cherkassky and Ma (2004). The noise variance is the remaining variance of the observations y, which cannot be explained by an ideal model. This ideal model has to be approximated with the nearest neighbor model (Cherkassky and Ma 2004), resulting in additional computational effort.

Range: `epsilon` $\in (0, \infty)$.

Transformation: `trans_10pow`.

Bounds: `lower = -8; upper = 0`.

Table 3.7 Survey of examples from the literature, for tuning of SVM

Hyperparameter	Lower bound	Upper bound	Result	Notes
(Probst et al. 2019a), various applications, 38 data sets				
kernel			radial basis	
cost	2^{-10}	2^{10}	0,002 to 963,81 *	
gamma	2^{-10}	2^{10}	0,003 to 276,02 *	
degree	2	5	2 to 4 *	
(Mantovani et al. 2015), various applications, 70 data sets				
cost	2^{-2}	2^{15}		
gamma	2^{-15}	2^{3}		
(van Rijn and Hutter 2018), various applications, 100 data sets				
cost	2^{-5}	2^{15}		
gamma	2^{-15}	2^{3}		
coef0	-1	1		Only sigmoid
tolerance	10^{-5}	10^{-1}		
(Sudheer et al. 2013), flow rate prediction (hydrology), 1 data set				
cost	10^{-5}	10^{5}	1,12 to 1,93 *	
epsilon	0	10	0,023 to 0,983 *	
gamma	0	10	0,59 to 0,87 *	

*Denotes that results depend on data set (multiple data sets)

Constraints: none.
Interactions: none are known.

In conclusion, Table 3.7 provides a brief survey of examples from the literature, where SVM was tuned.

3.8 Deep Neural Networks

3.8.1 Description

While DL describes the methodology, Deep Neural Networks (DNNs) are the models used in DL. DL models require the specification of a set of architecture-level parameters, which are important *hyperparameters*. Hyperparameters in DL are optimized in the outer loop of the hyperparameter tuning process. They are to be distinguished from the *parameters* of the DL method that are optimized in the initial loop, e.g., during the training phase of a Neural Network (NN) via backpropagation. Hyperparameter values are determined before the model is executed—they remain constant during model building and execution whereas parameters are modified. Selecting the method for the parameter optimization is a typical Hyperparameter Tuning (HPT) task. Available optimization methods such as ADAptive Moment estimation algorithm (ADAM) are described in Sect. 3.8.2.

Typical questions regarding hyperparameters in DL models are as follows:

1. How many layers should be combined?
2. Which dropout rate prevents overfitting?
3. How many filters (units) should be used in each layer?

Several empirical studies and benchmarking suites are available, see Sect. 6.2. But to date, there is no comprehensive theory that adequately explains how to answer these questions. Recently, Roberts et al. (2021) presented a first attempt to develop a DL theory.

Besides the hyperparameters discussed in this section, there are additional parameters used to define weight initialization schemes or regularization penalties. Furthermore, it should be noted that hyperparameters in DL methods can be conditionally dependent (this is also true for ML), e.g., on the number of layers as shown in the following example:

Example: Conditionally Dependent Hyperparameters

Mendoza et al. (2019) consider besides NN hyperparameters (e.g., batch size, number of layers, learning rate, dropout output rate, and optimizer), hyperparameters conditioned on solver type (e.g., β_1 and β_2) as well as hyperparameters conditioned on learning-rate policy, and per-layer hyperparameters (e.g., activation function, number of units). For practical reasons, Mendoza et al. (2019) constrained the number of layers to the range between one and six: firstly, they aimed to keep the training time of a single configuration low, and secondly each layer adds eight per-layer hyperparameters to the configuration space, such that allowing additional layers would further complicate the configuration process.

3.8.2   Hyperparameters of Deep Neural Networks

DL Hyperparameter `layers`

The parameter `layer` determines the number of layers of the NN. Only the number of hidden layers are affected, because input and output layers are basic elements of every NN. Larger values mean more complex models, which correspondingly also have more model coefficients, a higher runtime, but possibly also a higher model quality. There is also an increased risk of overfitting, if no regularization measures are implemented or methods such as early-stopping be used (Prechelt 2012).

Type: integer.
Default: 1.
Sensitivity: The influence of layers can be extreme. By varying this value,
                         extremely simple (no hidden layer or only one hidden layer with
                         very few neurons) or extremely complex models (thousands of
                         layers and neurons) can be generated. Moreover, the study of
                         Li et al. (2018) shows that network depth has a strong influ-
                         ence on weight optimization. The functional relationship between
                         weights and model quality becomes increasingly nonlinear as net-
                         work depth increases and contains more local optima. Thus, the
                         difficulty of weight optimization problem increases. At the same
                         time, this difficulty decreases when more neurons are used per
                         layer (Li et al. 2018). Also, "skip connections" (connections in
                         the network that skip layers) can help reduce the difficulty.
Heuristics: We are not aware of any quantitative heuristics. Bengio (2012)
                         recommend choosing the number of layers as large as possible,
                         considering the impact on computational resources. Larger net-
                         works exhibit better model performance as long as appropriate
                         regularization procedures are applied (Bengio 2012).
Range: $layers_i \in [1, \infty)$, with $i = \{1, 2, \ldots, \infty\}$. Only integer values
                         are valid.
Transformation: identity.
Bounds: lower = 1; upper = 4.
Constraints: none.
Interactions: An interaction of units and dropout with layers is
                         expected. These parameters together determine the total num-
                         ber of nodes in the network. This is also shown by the example
                         of Srivastava et al. (2014).

DL Hyperparameter units

The parameter units determines the size of the corresponding network layer (number of neurons in the layer). Only the hidden layers are affected, because the dimension of the input and output layers is pre-determined, i.e., the number of units of
the input layer depends on the dimensionality of the data and the number of units
of the output layer depends on the task (e.g., binary and multi-class classification
or regression). Similar to the layers, larger values mean more complex models,
which correspondingly also have more model coefficients, a higher runtime, but
possibly also a higher model quality. There is also an increased risk of overfitting,
should no regularization measures be taken or methods such as early-stopping be
used (Prechelt 2012).

Type:	integer, vector.
Default:	5.
Sensitivity:	The influence of `units` can be extreme. By varying this vector, extremely simple (no hidden layer or only one hidden layer with very few neurons) or extremely complex models (thousands of layers and neurons) can be generated. Moreover, the study of Li et al. (2018) shows that network depth has a strong influence on weight optimization. The functional relationship between weights and model quality becomes increasingly nonlinear as network depth increases and contains more local optima. Thus, the difficulty of weight optimization problem increases. At the same time, this difficulty decreases when more neurons are used per layer (Li et al. 2018). Also, "skip connections" (connections in the network that skip layers) can help reduce the difficulty.
Heuristics:	We are not aware of any quantitative heuristics. Larger networks exhibit better model performance as long as appropriate regularization procedures are applied (Bengio 2012). In addition, it is recommended from empirical results (Bengio 2012), to choose a first hidden layer that has more neurons than the input layer (i.e., the first element of `units` should be larger than n).
Range:	$units_i \in [1, \infty)$, with $i = \{1, 2, \ldots, \infty\}$. Only integer values are valid.
Transformation:	`trans2_pow`
Bounds:	`lower = 0; upper = 5`
Constraints:	none.
Interactions:	An interaction of `layers` and `dropout` with `units` is expected. These parameters together determine the total number of nodes in the network. This is also shown by the example of Srivastava et al. (2014).

DL Hyperparameter `activation`

The parameter `activation` specifies the activation function of the network nodes (neurons). In tensorflow, this parameter is often specified for each layer. This function decides how the input values of each node are translated into an output value.

The choice of activation function can have a strong impact on model performance. Among other things, `activation` influences an essential property of the network: the ability to approximate nonlinear functions. Only nonlinear activation functions allow this (Goldberg 2016).

Type:	character/function, vector. Standard activation functions can be selected via their name, else custom functions can be implemented in tensorflow or keras.

Default:	`relu` (parameter is not tuned).
Sensitivity:	unknown.
Heuristics:	A heuristic is not known. A popular choice is `activation =` `relu` (Bengio 2012). However, `activation` = tanh also shows success (LeCun et al. 2012). The choice of activation function is often empirically justified (Goldberg 2016), based on empirical data or empirical research for a specific problem. This underscores the need to tune this parameter.
Range:	`activation` ∈ {`tanh`, `sigmoid`, `relu`, `linear`, `swish`, ...}.
Transformation:	not implemented, because parameter is not tuned.
Bounds:	not implemented, because parameter is not tuned.
Constraints:	As a soft constraint, the choice of activation function may affect whether or not GPU-acceleration can be used in tensorflow. That is, some activation functions cannot be used if GPU support is required.
Interactions:	not known.

DL Hyperparameter `dropout`

Dropout is a commonly used regularization technique for DNNs: some percentage of the layer's output features will be randomly set to zero ("dropped out") during training, i.e., dropout refers to the random removal of nodes (units) in the network (Chollet and Allaire 2018; Srivastava et al. 2014). The parameter `dropout` (often also p (Srivastava et al. 2014)) is the probability that any node will be removed. Removing nodes randomly helps to avoid overfitting, `dropout` thus acts in the sense of regularization (Srivastava et al. 2014). In tensorflow, this parameter is often specified for each layer.

Type:	double, vector.
Default:	0.
Sensitivity:	A NN model's quality can be very sensitive to `dropout`. In an example, Srivastava et al. (2014) show that at a constant number of hidden nodes (network structure remains unchanged) the model error on test data for values between `dropout` = 0.4 and `dropout` = 0.6 is approximately constant. However, the model error increases for larger and smaller values of `dropout`.
Heuristics:	none known.
Range:	`dropout` ∈ (0, 1].
Transformation:	`identity`.
Bounds:	`lower = 0; upper = 0.4`.
Constraints:	none.

Interactions: An interaction of dropout with units and layers is
                         expected. These parameters together determine the total number
                         of nodes in the network. This is also illustrated in the example of
                         Srivastava et al. (2014).

DL Hyperparameter `learning_rate`

The learning rate (learning_rate) is a parameter of the weight optimization
algorithm employed in the NN. It can be understood as a multiplier for the gradient
in each iteration of the NN training procedure. The result is used to determine new
values for the network weights (Bengio 2012).

The learning rate is essential to the model. When the gradient of the weights is
determined, the learning rate decides how large a step to take in the direction of the
gradient. Very large values can lead to faster progress on the one hand, but on the
other hand can lead to instability and thus prevent the convergence of the training.

Type: double, scalar/vector. Usually a scalar, but a schedule of different
                         values can also be supplied to most tensorflow optimizers.
Default: 1e-3.
Sensitivity: Learning rates have a significant impact on the model. Accord-
                         ing to Bengio (2012), this parameter is often the most important
                         parameter that should always be considered when tuning neural
                         networks.
Heuristics: LeCun et al. (2012) propose to estimate learning rates individually
                         for each weight, proportional to the root of the number of inputs
                         to a node. Bengio (2012), on the other hand, states "The optimal
                         learning rate is usually close to (by a factor of 2) the largest learn-
                         ing rate that does not cause divergence of the training criterion."
                         Heuristics based on this observation require multiple restarts of
                         network training procedure (for example, start with large learn-
                         ing rate, stepwise divide by three until model training starts to
                         converge (Bengio 2012).)
Range: learning_rate $\in (0, \infty)$.
Transformation: identity.
Bounds: lower = 1e-6; upper = 1e-2.
Constraints: none.
Interactions: An interaction of batch_size, epochs, and learning_
                         rate is expected: Smaller learning rates or batch sizes may result
                         in larger epochs being required for model convergence.

DL Hyperparameter `epochs`

The parameter `epochs` determines the number of iterations (here: epochs), which are executed during the training of the model. An epoch describes the update of the network weights based on the calculated local gradient. Usually, within an epoch, the entire training data set is considered for determining the gradient (Bengio 2012). Each epoch can be subdivided again (depending on `batch_size`) into single steps.

In practice, `epochs` is often not a classical tuning parameter, since it mainly affects the runtime of the tuning procedure. Larger values are generally better for the model quality, but detrimental for the required runtime. However, larger runtimes may also increase the risk of overfitting, if no countermeasures are employed.

Type:	integer, scalar.
Default:	4.
Sensitivity:	For small values of `epochs`, the NN is sensitive to changes in `epochs`. It becomes increasingly insensitive to changes as `epochs` increases (i.e., as the model increasingly converges).
Heuristics:	None known.
Range:	`epochs` $\in [1, \infty]$. Only integer values are valid.
Transformation:	`trans_2pow`
Bounds:	`lower = 3; upper = 7`.
Constraints:	none.
Interactions:	See `batch_size` and `learning_rate`.

DL Hyperparameter `optimizer`

Optimization algorithms, e.g., Root Mean Square Propagation (RMSProp) (implemented in Keras as `optimizer_rmsprop`) or ADAM (`optimizer_adam`). Choi et al. (2019) considered RMSProp with momentum (Tieleman and Hinton 2012), ADAM (Kingma and Ba 2015), and ADAM (Dozat 2016) and claimed that the following relations hold:

$$SGD \subseteq MOMENTUM \subseteq RMSPROP,$$
$$SGD \subseteq MOMENTUM \subseteq ADAM,$$
$$SGD \subseteq NESTEROV \subseteq NADAM.$$

ADAM can approximately simulate MOMENTUM: MOMENTUM can be approximated with ADAM, if a learning-rate schedule that accounts for ADAM's bias correction is implemented. Choi et al. (2019) demonstrated that these inclusion relationships are meaningful in practice. In the context of HPT and Hyperparameter Optimization (HPO), inclusion relations can significantly reduce the complexity of the experimental design. These inclusion relations justify the selection of a basic set, e.g., RMSProp, ADAM, and Nesterov-accelerated Adaptive Moment Estimation (NADAM).

Type:	factor.
Default:	5.
Sensitivity:	unknown.
Heuristics:	We are not aware of heuristics.
Range:	`optimizer` $\in$ { `"SDG"`, `RMSPROP"`, `ADAGRAD"`, `ADADELTA"`, `ADAM"`, `ADAMAX"`, `NADAM"` }.
Transformation:	`identity`.
Bounds:	`lower = 1; upper = 7`.
Constraints:	none.
Interactions:	Necessarily, there is an interaction.

DL Hyperparameter `loss`

This parameter determines the loss function that is minimized when training the network (optimizing the weights). The loss function can have a significant influence on the quality of the model (Janocha and Czarnecki 2017). However, it is not a typical tuning parameter, in part because the tuning procedure itself requires a consistent loss function, to identify better configurations of the hyperparameters. The `loss` parameter is therefore usually chosen separately by the user before the tuning procedure.

Type:	character, scalar.
Default:	problem dependent, parameter is not tuned.
Sensitivity:	not known.
Heuristics:	not known.
Range:	several standard loss functions (such as Mean Squared Error (MSE)) are available in tensorflow, custom loss functions can be provided by users.
Transformation:	not implemented, because parameter is not tuned.
Bounds:	not implemented, because parameter is not tuned.
Constraints:	Some loss functions are specific to certain tasks (i.e., classification: crossentropy, regression: MSE).
Interactions:	unknown.

DL Hyperparameter `batch_size`

When determining the gradient of the network weights, either the whole data set can be used for this or only a subset (here: batch). The size of this subset is specified by `batch_size`.

The parameter `batch_size` mainly affects the runtime of the training (Bengio 2012). However, `batch_size` also affects the quality of the model. Small batch sizes may introduce a strong random element to weigh updates, which can hinder or benefit the learning process. Shallue et al. (2019) and Zhang et al. (2019) have shown empirically that increasing the batch size can increase the gaps between training times for different optimizers.

Type:	integer, scalar.
Default:	32.
Sensitivity:	unknown.
Heuristics:	We are not aware of heuristics, 32 is suggested as a good default value (Bengio 2012). However, from the experience of the authors of this expertise, this is highly dependent on the data situation, computer architecture, and further configuration of the model. Specifying `batch_size` as a function of n should also be considered.
Range:	`batch_size` $\in (1, n]$. Only integer values are valid. Common `batch_size` values are between 10 and several hundred (Bengio 2012). But several thousands are also possible (Mendoza et al. 2016).
Transformation:	not implemented, because parameter is not tuned.
Bounds:	not implemented, because parameter is not tuned.
Constraints:	none.
Interactions:	Necessarily, there is an interaction between `batch_size` and `epochs`, since both together determine the number of steps of the training procedure. In addition, an interaction of `batch_size`, `epochs`, and `learning_rate` is also expected. The interaction between `batch_size` and `learning_rate` is also mentioned by Bengio (2012).

3.9 Summary and Discussion

On the basis of our literature survey, we recommend tuning the introduced hyperparameters of ML models. In the experiments described in this study, we also investigate five additional parameters:

- `dropoutfact` is a multiplier for `dropout`, which reduces or increases `dropout` in each consecutive layer of the network;
- `unitsfact` performs the same job but for `units`; and
- `beta_1`, `beta_2`, and `epsilon` are parameters affecting the `optimizer`.

Reasonable bounds for all investigated parameters are summarized in Table 3.8.

Table 3.8 Overview of hyperparameters in the experiments. For data type, we employ the signifiers used in R. For categorical parameters, we list categories instead of providing bounds. "Default" refers to the ML default values in mlr and to the DL default values in SPOTMisc

Model	Hyperparameter	Data type	Default (trans.)	Lower bound (trans.)	Upper bound (trans.)	Lower bound	Upper bound	Transformation
KNN	k	Integer	7	1	30	1	30	–
EN	p	Numeric	2	0.1	100	–1	2	10^p
	alpha	Numeric	1	0	1	0	1	–
	thresh	Numeric	10^{-7}	10^{-8}	10^{-1}	–8	–1	10^{thresh}
DT	minsplit	Integer	20	1	300	1	300	–
	minbucket	Integer	20/3	1	150	0.1	0.5	round(max(minsplit × minbucket, 1))
	cp	Numeric	10^{-2}	10^{-10}	10^0	–10	0	10^{cp}
	maxdepth	Integer	30	1	30	1	30	–
RF	num.trees	Integer	500	1	2048	0	11	round($2^{num.trees}$)
	mtry	Integer	floor($\sqrt{n}$)	1	n	1	n	–
	sample.fraction	Numeric	1	0.1	1	0.1	1	–
	replace	Factor	TRUE	TRUE, FALSE				–
	respect.unordered.factors	Factor	ignore	ignore, order				–

(continued)

Table 3.8 (continued)

Model	Hyperparameter	Data type	Default (trans.)	Lower bound (trans.)	Upper bound (trans.)	Lower bound	Upper bound	Transformation
xgBoost	eta	Numeric	0.3	2^{-10}	2^{0}	-10	0	2^{eta}
	nrounds	Integer	1	1	2048	0	11	$2^{nrounds}$
	lambda	Numeric	1	2^{-10}	2^{10}	-10	10	2^{lambda}
	alpha	Numeric	2^{-10}	2^{-10}	2^{10}	-10	10	2^{alpha}
	subsample	Numeric	1	0.1	1	0.1	1	–
	colsample_ bytree	Numeric	1	$1/n$	1	$1/n$	1	–
	gamma	Numeric	2^{-10}	2^{-10}	2^{10}	-10	10	2^{gamma}
	maxdepth x	Integer	6	1	15	1	15	–
	min_child_ weight	Numeric	1	1	128	0	7	$2^{min\_child\_weight}$
SVM	kernel	Factor	radial	radial, sigmoid				–
	gamma	Numeric	$1/n$	2^{-10}	2^{10}	-10	10	2^{gamma}
	coef0	Numeric	0	-1	1	-1	1	–
	cost	Numeric	1	2^{-10}	2^{10}	-10	10	2^{cost}
	epsilon	Numeric	0.1	10^{-8}	10^{0}	-8	0	$10^{epsilon}$
DL	dropout	Numeric	0	0	0.4	0	0.4	–
	dropoutfact	Numeric	0	0	0.5	0	0.5	–
	units	Integer	32	1	32	0	5	2^{units}
	unitsfact	Numeric	0.5	0.25	1	0.25	1	–
	learning_rate	Numeric	1e-3	1e-6	1e-2	1e-6	1e-2	–
	epochs	Integer	16	8	128	3	7	2^{epochs}
	beta_1	Numeric	0.9	0.9	0.99	0.9	0.99	–
	beta_2	Numeric	0.999	0.99	0.9999	0.99	0.9999	–
	layers	Integer	1	1	4	1	4	–
	epsilon	Numeric	1e-7	1e-9	1e-8	1e-9	1e-8	–
	optimizer	Factor	5	1	7	1	7	–

Chapter 4
Hyperparameter Tuning Approaches

Thomas Bartz-Beielstein and Martin Zaefferer

Abstract This chapter provides a broad overview over the different hyperparameter tunings. It details the process of HPT, and discusses popular HPT approaches and difficulties. It focuses on surrogate optimization, because this is the most powerful approach. It introduces Sequential Parameter Optimization Toolbox (SPOT) as one typical surrogate method. SPOT is well established and maintained, open source, available on Comprehensive R Archive Network (CRAN), and catches mistakes. Because SPOT is open source and well documented, the human remains in the loop of decision-making. The introduction of SPOT is accompanied by detailed descriptions of the implementation and program code. This chapter particularly provides a deep insight in Kriging (aka Gaussian Process (GP) aka Bayesian Optimization (BO)) as a workhorse of this methodology. Thus it is very hands-on and practical.

4.1 Hyperparameter Tuning: Approaches and Goals

The following HPT approaches are popular:

- manual search (or trial-and-error (Meignan et al. 2015)),
- simple Random Search (RS), i.e., randomly and repeatedly choosing hyperparameters to evaluate,
- grid search (Tatsis and Parsopoulos 2016),
- directed, model free algorithms, i.e., algorithms that do not explicitly make use of a model, e.g., Evolution Strategys (ESs) (Hansen 2006; Bartz-Beielstein et al. 2014) or pattern search (Lewis et al. 2000),

T. Bartz-Beielstein (✉)
Institute for Data Science, Engineering and Analytics, TH Köln, Gummersbach, Germany
e-mail: thomas.bartz-beielstein@th-koeln.de

M. Zaefferer
Bartz & Bartz GmbH and with Institute for Data Science, Engineering, and Analytics, TH Köln, Gummersbach, Germany

Duale Hochschule Baden-Württemberg Ravensburg, Ravensburg, Germany
e-mail: zaefferer@dhbw-ravensburg.de

E. Bartz et al. (eds.), *Hyperparameter Tuning for Machine and Deep Learning with R*,
https://doi.org/10.1007/978-981-19-5170-1_4

71

- hyperband, i.e., a multi-armed bandit strategy that dynamically allocates resources to a set of random configurations and uses successive halving to stop poorly performing configurations (Li et al. 2016),
- Surrogate Model Based Optimization (SMBO) such as SPOT, (Bartz-Beielstein et al. 2005, 2021).[1]

Manual search and grid search are probably the most popular algorithms for HPT. Similar to suggestions made by Bartz-Beielstein et al. (2020a), we propose the following recommendations for performing HPT studies:

(R-1) Goals: clearly state the reasons for performing HPT. Improving an existing solution, finding a solution for a new, unknown problem, or benchmarking two methods are only three examples with different goals. Each of these goals requires a different experimental design.

(R-2) Problems: select suitable problems. Decide, how many different problems or problem instances are necessary. In some situations, surrogates (e.g., Computational Fluid Dynamics (CFD) simulations) can accelerate the tuning (Bartz-Beielstein et al. 2018).

(R-3) Algorithms: select a portfolio of ML and DL algorithms to be included in the HPT experimental study. Consider base-line methods such as RS and methods with their default hyperparameter settings.

(R-4) Performance: specify the performance measure(s). See the discussion in Sect. 2.2.

(R-5) Analysis: describe how the results can be evaluated. Decide, whether parametric or non-parametric methods are applicable. See the discussion in Chap. 5.

(R-6) Design: set up the experimental design of the study, e.g., how many runs shall be performed. Tools from Design of Experiments (DOE) and Design and Analysis of Computer Experiments (DACE) are highly recommended. See the discussion in Sect. 5.6.5.

(R-7) Presentation: select an adequate presentation of the results. Consider the audience: a presentation for the management might differ from a publication in a journal.

(R-8) Reproducibility: consider how to guarantee scientifically sound results and how to guarantee a lasting impact, e.g., in terms of comparability. López-Ibáñez et al. (2021a) present important ideas.

In addition to these recommendations, there are some specific issues that are caused by the ML and DL setup.

We consider a HPT approach based on SPOT that focuses on the following topics:

Limited Resources.   We focus on situations, where limited computational resources are available. This may be simply due to the availability and

---

[1] The acronym SMBO originated in the engineering domain (Booker et al. 1999; Mack et al. 2007). It is also popular in the ML community, where it stands for *sequential model-based optimization*. We will use the terms *sequential model-based optimization* and *surrrogate model-based optimization* synonymously.

cost of hardware, or because confidential data has to be processed strictly locally.

Understanding.   In contrast to standard HPO approaches, SPOT provides statistical tools for *understanding* hyperparameter importance and interactions between several hyperparameters.

Explainability.   Understanding is a key tool for enabling transparency and explainability, e.g., quantifying the contribution of ML and DL components (layers, activation functions, etc.).

Replicablity.   The software code used in this study is available in the open source R software environment for statistical computing and graphics (R) package SPOT via the CRAN. Replicability is discussed in Sect. 2.7.2. SPOT is a well-established open-source software, maintained for more than 15 years (Bartz-Beielstein et al. 2005).

Furthermore, Falkner et al. (2018) claim that practical HPO solutions should fulfill the following requirements:

- strong anytime and final performance,
- effective use of parallel resources,
- scalability, as well as robustness and flexibility.

For sure, we are not seeking the overall best hyperparameter configuration that results in a method which outperforms any other method in every problem domain (Wolpert and Macready 1997). Results are specific for one problem instance—their generalizability to other problem instances or even other problem domains is not self-evident and has to be proven (Haftka 2016).

4.2 Special Case: Monotonous Hyperparameters

A special case is hyperparameters with monotonous effect on the quality and run time (and/or memory requirements) of the tuned model. In our survey (see Table 4.1), two examples are included: `num.trees` (RF) and `thresh` (EN). Due to the monotonicity properties, treating these parameters differently is a likely consideration. In the following, we focus the discussion on `num.trees` as an example, since this parameter is frequently discussed in literature and online communities (Probst et al. 2018).

It is known from the literature that larger values of `num.trees` generally lead to better models. As the size increases, a saturation sets in, leading to progressively lower quality gains. It should be noted that this is not necessarily true for every quality measure. Probst et al. (2018), for example, show that this relation holds for log-loss and Brier score, but not for Area Under the receiver operating characteristic Curve (AUC).

Table 4.1 Global hyperparameter overview. The column "Quality" shows all parameter, where a monotonous relationship between parameter values and model quality is to be expected. (↑↑: quality increases if parameter value increases, ↑↓: quality decreases if parameter value increases). Correspondingly, the column "run time" shows the same information for the relationship of parameter values and run time

Model	Hyperparameter	Quality	Run time
KNN	k		↑↑
	p		
EN	alpha		
	lambda		
	thresh	↑↓	↑↓
DT	minsplit		↑↓
	minbucket		↑↓
	cp		↑↓
	maxdepth		↑↑
RF	num.trees	↑↑	↑↑
	mtry		↑↑
	sample.fraction		↑↑
	replace		
	respect.unordered.factors		
xgBoost	eta		
	nrounds		↑↑
	lambda		
	alpha		
	subsample		↑↑
	colsample_bytree		↑↑
	gamma		↑↓
	max_depth		↑↑
	min_child_weight		↑↓
SVM	kernel		
	degree		
	gamma		
	coef0		
	cost		
	epsilon		

Because of this relationship, Probst et al. (2018) claim that num.trees should not be optimized. Instead, it is recommended setting the parameter to a "computationally feasible large number" (Probst et al. 2018). For certain applications, especially for relatively small or medium-sized data sets, we support this assessment. However, at least in perspective, the analysis in this book considers tuning hyperparameters for very large data sets (many observations and/or many features). For this use case, we do not share this recommendation, because the required run time of the model plays an increasingly important role and is not explicitly considered in the recommendation. In this case, a "computationally feasible large number" is not trivial to determine.

In total, we consider five solutions for handling monotonous hyperparameters:

(M-1) Set manually: The parameter is set to the largest possible value that is still feasible with the available computing resources. This solution involves the following risks:

    a. Single evaluations during tuning waste time unnecessarily.
    b. Interactions with parameters (e.g., `mtry`) are not considered.
    c. The value may be unnecessarily large (from a model quality point of view).
    d. The determination of this value can be difficult, it requires detailed knowledge regarding: size of the data set, efficiency of the model implementation, available resources (memory/computer cores / time).

(M-2) Manual adjustment of the tuning: After a preliminary examination (as represented, e.g., by the initial design step of SPOT) a user intervention takes place. Based on the preliminary investigation, a value that seems reasonable is chosen by the user and is not changed in the further course of the tuning. This solution involves the following risks:

    a. The preliminary investigation itself takes too much time.
    b. The decision after the preliminary investigation requires intervention by the user (problematic for automation). While this is feasible for individual cases, it is not practical for numerous experiments with different data (as in the experiments of the study in Chap. 12). Moreover, this reduces the reproducibility of the results.
    c. Depending on the scope and approach of the preliminary study, interactions with other parameters may not be adequately accounted for.

(M-3) No distinction: parameters like `num.trees` are optimized by the tuning procedure just like all other hyperparameters. This solution involves the following risks:

    a. The upper bound for the parameter is set too low, so potentially good models are not explored by the tuning procedure. (Note: bounds set too tight for the search space are a general risk that can affect all other hyperparameters as well).
    b. The upper bound is set too high, causing individual evaluations to use unnecessary amounts of time during tuning.
    c. The best found value may become unnecessarily large (from a model quality point of view).

(M-4) Multi-objective: run time and model quality can be optimized simultaneously in the context of multi-objective optimization. This solution involves the following risks:

    a. Again, manual evaluation is necessary (selection of a sector of the Pareto front) to avoid that from a practical point of view irrelevant (but possibly Pareto-optimal) solutions are investigated.

    b. This manual evaluation also reduces reproducibility.

(M-5) Regularization via weighted sum: The number of trees (or similar parameters) can be incorporated into the objective function. In this case, the objective function becomes a weighted sum of model quality and number of trees (or run time), with a weighting factor θ.

    a. The new parameter of the tuning procedure, θ, has to be determined.

    b. Moreover, the optimization of a weighted sum cannot find certain Pareto-optimal solutions if the Pareto front is non-convex.

In the experimental investigation in Chap. 12, we use solution (M-3). That is, the corresponding parameters are tuned but do not undergo any special treatment during tuning. Due to the large number of experiments, user interventions would not be possible and would also complicate the reproducibility of the results. In principle, we recommend this solution for use in practice.

In individual cases, or if a good understanding of algorithms and data is available, solution (M-2) can also be used. For this, SPOT can be interrupted after the first evaluation step, in order to set the corresponding parameters to a certain value or to adjust the bounds if necessary (e.g., if num.trees was examined with too low an upper bound).

4.3 Model-Free Search

4.3.1 Manual Search

A frequently applied approach is that ML and DL methods are configured manually (Bergstra and Bengio 2012). Users apply their own experience and trial-and-error to find reasonable hyperparameter values.

In individual cases, this approach may indeed yield good results: when expert knowledge about data, methods, and parameters is available. At the same time, this approach has major weaknesses, e.g., it may require significant amount of work time by the users, bias may be introduced due to wrong assumptions, limited options for parallel computation, and extremely limited reproducibility. Hence, an automated approach is of interest.

4.3.2 Undirected Search

Undirected search algorithms determine new hyperparameter values independently of any results of their evaluation. Two important examples are Grid Search and RS.

Grid Search covers the search space with a regular grid. Each grid point is evaluated. RS selects new values at random (usually independently, uniform distributed) in the search space.

Grid Search is a frequently used approach, as it is easy to understand and implement (including parallelization). As discussed by Bergstra and Bengio (2012), RS shares the advantages of Grid Search. However, they show that RS may be preferable to Grid Search, especially in high-dimensional spaces or when the importance of individual parameters is fairly heterogeneous. They hence suggest to use RS instead Grid Search if such simple procedures are required. Probst et al. (2019a) also use a RS variant to determine the tunability of models and hyperparameters. For these reasons, we employ RS as a baseline for the comparison in our experimental investigation in Chap. 12.

Next to Grid Search and RS, there are other undirected search methods. Hyperband is an extension of RS, which controls the use of certain resources (e.g., iterations, training time) (Li et al. 2018). Another relevant set of methods is the Design of Experiments methods, such as Latin Hypercube Designs (Leary et al. 2003).

! Attention: Random Search Versus Grid Search

Interestingly, Bergstra and Bengio (2012) demonstrate empirically and show theoretically that randomly chosen trials are more efficient for HPT than trials on a grid. Because their results are of practical relevance, they are briefly summarized here: In grid search the set of trials is formed by using every possible combination of values, grid search suffers from the curse of dimensionality because the number of joint values grows exponentially with the number of hyperparameters.

> A Gaussian process analysis of the function from hyper-parameters to validation set performance reveals that for most data sets only a few of the hyper-parameters really matter, but that different hyper-parameters are important on different data sets. This phenomenon makes grid search a poor choice for configuring algorithms for new data sets (Bergstra and Bengio 2012).

Let Ψ denote the space of hyperparameter response functions. Bergstra and Bengio (2012) claim that RS is more efficient in ML than grid search because a hyperparameter response function $\psi \in \Psi$ usually has a low effective dimensionality (see Definition 2.25), i.e., ψ is more sensitive to changes in some dimensions than others (Caflisch et al. 1997).

The observation that only a few of the parameters matter can also be made in the engineering domain, where parameters such as pressure or temperature play a dominant role. In contrast to DL, this set of important parameters does not change fundamentally in different situations. We assume that the high variance in the set of important DL hyperparameters is caused by confounding.

Due to its simplicity, it turns out in many situations that RS is the best solution, especially in high-dimensional spaces. Hyperband should also be mentioned in this context, although it can result in a worse final performance than model-based approaches, because it only samples configurations randomly and does not learn from previously sampled configurations (Li et al. 2016). Bergstra and Bengio (2012) note that RS can probably be improved by automating what manual search does, i.e., using SMBO approaches such as SPOT.

HPT is a powerful technique that is an absolute requirement to get to state-of-the-art models on any real-world learning task, e.g., classification and regression. However, there are important issues to keep in mind when doing HPT: for example, validation-set overfitting can occur, because hyperparameters are usually optimized based on information derived from the validation data.

4.3.3 Directed Search

One obvious disadvantage of undirected search is that a large amount of the computational effort may be spent on evaluating solutions that cover the whole search space. Hence, only a comparatively small amount of the computational budget will be spent on potentially optimal or at least promising regions of the search space.

Directed search on the other hand may provide a more purposeful approach. Basically any gradient-free, global optimization algorithm could be employed. Prominent examples are Iterative Local Search (ILS) (Hutter et al. 2010b) and Iterative Racing (IRACE) (López-Ibáñez et al. 2016). Metaheuristics like Evolutionary Algorithms (EAs) or Swarm Optimization are also applicable (Yang and Shami 2020). In comparison to undirected search procedures, directed search has two frequent drawbacks: an increased complexity that makes implementation a larger issue, and being more complicated to parallelize.

We employ a *model-based* directed search procedure in this book, which is described in the following Sect. 4.4.

4.4 Model-Based Search

A disadvantage of model-free, directed search procedures is that they may require a relatively large number of evaluations (i.e., long run times) to approximate the values of optimal hyperparameters.

Tuning ML and DL algorithms can become problematic if complex methods are tuned on large data sets, because the run time for evaluating a single hyperparameter configuration may go up into the range of hours or even days. Model-based search is one approach to resolve this issue. These search procedures use information gathered during the search to learn the relationship between hyperparameter values and performance measures (e.g., misclassification error). The model that encodes this learned relationship is called the surrogate model or *surrogate*.

Definition 4.1 (*Surrogate Optimization*) The surrogate optimization uses two phases.

Construct Surrogate   Generate (random) solutions. Evaluate the (expensive) objective function at these points. Construct a surrogate, S, of the objective function, e.g., by building a GP aka Kriging model (surrogate).

Search for Minimum   Search for a minimum of the objective function on the (cheap) surrogate. Choose the best point as a candidate. Evaluate the objective function at the best candidate point. This point is called an infill point. Update the surrogate using this value and search again.

The advantage of this surrogate optimization is that a considerable part of the evaluation burden (i.e., the computational effort) can be shifted from *real* evaluations to evaluations of the surrogate, which should be faster to evaluate.

In HPT, mixed optimization problems are common, i.e., the variables are continuous or discrete (Cuesta Ramirez et al. 2022). Bartz-Beielstein and Zaefferer (2017) provide an overview of metamodels that have or can be used in optimization. They show how it was made possible by the realization that GP kernels (covariance functions) in mixed variables can be created by composing continuous and discrete kernels. In this case, the infill criterion (acquisition function) is defined over the same space as the objective function. Therefore maximizing the acquisition function is also a mixed variables problem.

One variant of model-based search is SPOT (Bartz-Beielstein 2005), which be will described in Sect. 4.5.

4.5 Sequential Parameter Optimization Toolbox

SMBO methods are common approaches in simulation and optimization. SPOT has been developed, because there is a strong need for sound statistical analysis of simulation and optimization algorithms. SPOT includes methods for tuning based on classical regression and analysis of variance techniques; tree-based models such as Classification and Regression Trees (CART) and random forest; BO (Gaussian process models, aka Kriging), and combinations of different meta-modeling approaches.

Basic elements of the Kriging-based surrogate optimization such as interpolation, expected improvement, and regression are presented in the Appendix, see Sect. 4.6. The Sequential Parameter Optimization (SPO) toolbox implements a modified version of this method and will be described in this section.

SPOT implements key techniques such as exploratory fitness landscape analysis and sensitivity analysis. SPOT can be used for understanding the performance of algorithms and gaining insight into algorithm's behavior. Furthermore, SPOT can be used as an optimizer and for automatic and interactive tuning.

Details of SPOT and its application in practice are given by Bartz-Beielstein et al. (2021). SPOT was originally developed for the tuning of optimization algorithms. The requirements and challenges of algorithm tuning in optimization broadly reflect those of tuning machine learning models. SPOT uses the following approach (outer loop).

Setup: In a first step, several candidate solutions (here: different combinations of hyperparameter values) are created. These are steps (S-1) and (S-2) in the function `spot`, see Fig. 4.1.

Evaluate: All new candidate solutions are evaluated (here: training the respective ML or DL model with the specified hyperparameter values and measuring the quality / performance). This is step (S-3).

Termination: Check whether a termination criterion has been reached (e.g., number of iterations, evaluations, run time, or a satisfying solution has been found). These are steps (S-4) to (S-9).

Select: Samples for building the surrogate S are selected. This is step (S-10).

Training: The surrogate S will be trained with all data derived from the evaluated candidate solutions, thus learning how hyperparameters affect model quality. This is step (S-11).

Surrogate search: The trained model is used to perform a search for new, promising candidate solutions. These are steps (S-12) and (S-16).

Budget: Optimal Computing Budget Allocation (OCBA) is used to determine the number of repeated evaluations. This is step (S-17).

Evaluation The new solutions are evaluated on the objective function, e.g., the loss is determined. These are steps (S-18) to (S-22).

Exploit: An optimizer is used to perform a local search on S to refine the best solution found. These are steps (S-23) and (S-24). Whereas optimization on the surrogate in the main loop is a weighted combination of exploration and exploita-

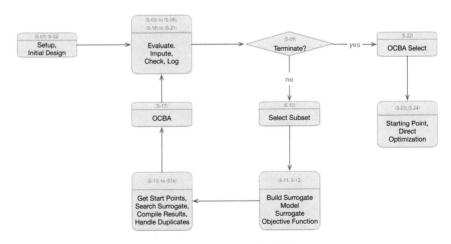

Fig. 4.1 Visual representation of model-based search with SPOT

tion, using Expected Improvement (EI) as a default weighting mechanism, this final optimization step is purely exploitative.

Note, that it can be useful to allow for user interaction with the tuner after the evaluation step. Thus, the user may affect changes of the search space (stretch or shrink bounds on parameters, eliminate parameters). However, we will consider an automatic search in our experiments.

We use the R implementation of SPOT, as provided by the R package SPOT (Bartz-Beielstein et al. 2021, 2021c). The SPOT workflow will be described in the following sections.

In the remainder of this book, SPOT will refer to the general method, whereas spot denotes the function from the R package SPOT.

Steps, subroutines and data of the spot process are shown in Fig. 4.2.

4.5.1 spot as an Optimizer

spot uses the same syntax as optim, R's general-purpose optimization based on Nelder-Mead, quasi-Newton, and conjugate-gradient algorithms (R Core Team 2022). spot can be called as shown in the following example.

Example: spot

SPOT comes with many pre-defined functions from optimization, e.g., Sphere, Rosenbrock, or Branin. These implementations use the prefix "fun", e.g., funSphere is the name of the sphere function. The package SPOTMisc provides funBBOBCall, an interface to the real-parameter Black-Box Optimization Benchmarking (BBOB) function suite (Mersmann et al. 2010a). Furthermore, users can also specify their own objective functions.

Searching for the optimum of the (two-dimensional) sphere function funSphere, i.e., $f(x) = \sum_{i=1}^{2} x_i^2$, on the interval between $(-1, -1)$ and $(1, 1)$ can be done as follows:

```
library("SPOT")
spot(x = NULL, fun = funSphere, lower = c(-1, -1), upper = c(1, 1))
```

Four arguments are passed to spot: no explicit starting point for the optimization is used, because the parameter x was set to NULL, the function funSphere, and the lower and upper bounds. The length of the lower bound argument defines the problem dimension n.

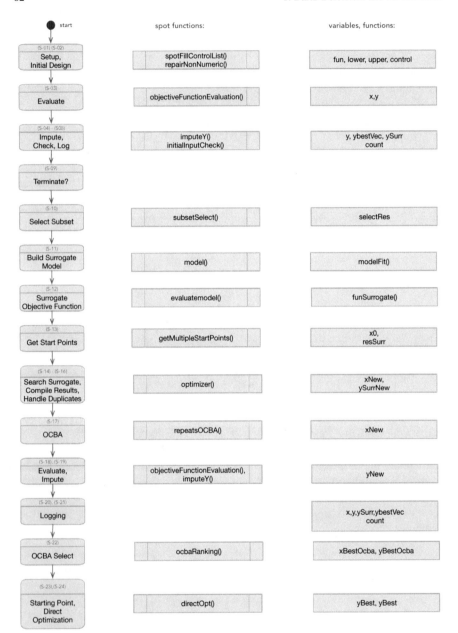

Fig. 4.2 Steps, functions and variables of the `spot` function

Table 4.2 SPOT parameters. This table shows the mandatory parameters. The list `control` can be used to pass additional parameters to `spot`. Additional arguments to the objective function `fun` can be passed via "...", similar to the varargs method in other programming languages

Parameter	Default value	Description
x	NULL	Starting point
fun		Objective function, e.g., `funSphere`, or as described in Sect. 8.44
lower		Lower bound, defines the problem dimension n
upper		Upper bound
control	List	See description in Table 4.3
...		Used to pass those additional arguments on to the objective function `fun`

> **Mandatory Parameters**

The arguments x, fun, lower, and upper are mandatory for spot, they are shown in Table 4.2.

Additional arguments can be passed to `spot`. They allow a very flexible handling, e.g., for passing extra arguments to the objective function `fun`. To improve the overview, parameters are organized as lists. The "main" list is called `control`, see Table 4.3. It collects `spot`'s parameters, some of them are organized as lists. They are shown in Table 4.4.

The `control` list is used for managing SPOT's parametrization, e.g., for defining hyperparameter types and ranges.

4.5.2 spot's Initial Phase

The initial phase consists of five steps (S-1) to (S-5). The corresponding R code is shown in Sect. 4.7.

(S-1) *Setup.* After performing an initial check on the control list, the `control` list is completed.

The `control` list contains the parameters from Table 4.3.

(S-2) *Initial design.* The parameter `seedSPOT` is used to set the seed for `spot` before the initial design X is generated. The design type is specified via `control$design`. The recommended design function is `designLHD`, i.e., a Latin Hypercube Design (LHD), which is also the default configuration.

Table 4.3 SPOT: parameters of the control list

Parameter	Default value, type	Description
design	designLHD, function	The design function is used to generate the initial design (see spot) and to generate multiple start points (see getMultiStartPoints)
directOpt	optimNLOPTR, function	Optimizer used for direct optimization after SMBO is done
funEvals	20	Number of objective function (fun) evaluations
infillCriterion	NULL, function	A function defining an infill criterion to be used while optimizing a model
model	buildKriging, function	A function that builds a statistical model of the observed data
multiStart	1 (no multi starts), integer	Number of restarts of the optimizer on the surrogate model
noise	FALSE, logical	
OCBA	paramFALSE, logical	Use OCBA
OCBABudget	3, integer	Budget for OCBA
optimizer	function	Optimizer on surrogate model
parNames	character, paste0 ("x", 1:dimension)	Hyperparameter names
plots	paramFALSE, logical	Show progress plots
progress	paramFALSE, logical	Show numerical information about the progress
replicateResults	paramFALSE, logical	Evaluate configuration(s), do not perform SMBO
replicates	integer	Number of replicates
returnFullControlList	logical	Return the full control list
seedFun	seed function for objective function	
seedSPOT	seed used for spot	
subsetSelect	selectAll	Subset used for fitting the surrogate model
tolerance	numerical	Sqrt(.Machine$double.eps)
transformFun	vector	Variable transformation
types	rep ("numeric", dimension)	Hyperparameter types
verbosity	integer	Verbosity
xNewActualSize	integer	Number of new design points proposed by the surrogate model
designControl	list	Parameters used by the design function
directOptControl	list	Parameters used by the direct function
modelControl	list	Parameters used by the surrogate model
optimizerControl	list	Parameters used by the optimizer
subsetControl	list	Parameters used by the subsetSelect function
time	list	Time related parameters
yImputation	list	List of functions to determine imputations, handleNAsMethod

Table 4.4 SPOT: parameters of the other lists

List	Parameter	Default value, type	Description
`designControl`	replicates	Rinit	
	size	initSizeFactor * length(cfg$lower)	
`directOptControl`	list	parameters used by the `direct` function. Direct optimization is performed, if `control$directOptControl$funEvals > 0`	
`modelControl`	target	krigingTarget	
	useLambda	krigingUseLambda	
	reinterpolate	krigingReinterpolate	re-interpolation is supposed to be used with stochastic experiments, which do need a non-interpolating model that avoids predicting nonzero error at sample locations. Can be useful when the model is deterministic, i.e., repeated evaluations of one parameter vector do not yield different values but does have a "noisy" structure (e.g., due to computational inaccuracies, systematical errors)
	infillCriterion	NULL	
`optimizerControl`	funEvals	multFun * length(cfg$lower))	
`time`	maxTime	Inf	time bugdet in minutes
`yImputation`	handleNAsMethod	handleNAsMethod	
	imputeCriteriaFuns	imputeCriteriaFuns	
	penaltyImputation	3	

Fig. 4.3 Initial design. The first ten points created by designLHD

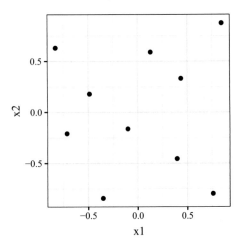

Ten initial design points are available now, because the default value of the parameter `designControl$size`, which specifies the initial design size, is set to 10 if the function `designLHD` is used (Fig. 4.3).

Program Code: Steps (S-1) and (S-2)

Steps (S-1) and (S-2) are implemented as follows:

```
## (S-1) Setup:
fun <- funNoise
lower <- c(-1, -1)
upper <- c(1, 1)
control <- list(
  OCBA = TRUE,
  OCBABudget = 3,
  replicates = 2,
  noise = TRUE,
  multiStart = 2,
  designControl = list(replicates = 2)
)
control <- spotFillControlList(control, lower, upper)
## (S-2) Initial design:
set.seed(control$seedSPOT)
x <- control$design(
  x = NULL,
  lower = lower,
  upper = upper,
  control = control$designControl
)
x <- repairNonNumeric(x, control$types)
```

Example: Modifying the initial design size

Arbitrary initial design sizes can be generated by modifying the `size` argument of the `designControl` list:

```
control$designControl$size <- 5
```

Here is the full code for starting `spot` with an initial design of size five:

```
spot (
  x = NULL,
  fun = funSphere,
  lower = c(-1, -1), upper = c(1, 1),
  control = list(designControl = list(size = 5))
)
```

Because the `lower` bound was set to $(-1, -1)$, a two-dimensional problem is defined, i.e., $f(x_1, x_2) = x_1^2 + x_2^2$. The result from this `spot` run is stored as a list in the variable `return`.

Variable types are assumed to be `numeric`, which is the default type if no other type is specified. Type information, which is available from `config$types`, is used to transform the variables. The function `spot` can handle the data types `numeric`, `integer`, and `factor`. The function `repairNonNumeric` maps non-numerical values to `integers`.

(S-3) *Evaluation of the Initial Design.* Using `objectiveFunction Evaluation`, the objective function `fun` is evaluated on the initial design matrix x.

In addition to `xnew`, a matrix of already known solutions, to determine whether Random Number Generator (RNG) seeds for new solutions need to be incremented, can be passed to the function `objectiveFunction Evaluation`.

! Transformation of Variables

If variable transformation functions are defined, the function `transformX` is applied to the parameters during the execution of the function `objectiveFunctionEvaluation`.

The function `objectiveFunctionEvaluation` returns the matrix y.

(S-4) *Imputation: Handling Missing Values.* The feasibility of the y-matrix is checked. Methods to handle `NA` and infinite y-values are applied, which are available via the function `imputeY`.

The spot loop starts after the initial phase. The function `spotLoop` is called.

Program Code: Steps (S-3) and (S-4)

Steps (S-3) and (S-4) are implemented as follows:

```
## (S-3) Eval initial design:
y <- objectiveFunctionEvaluation(
  x = NULL,
  xnew = x,
  fun = fun,
  control = control
)
## (S-4) Imputation:
if (!is.null(control$yImputation$handleNAsMethod)) {
  y <- imputeY(
    x = x,
    y = y,
    control = control
  )
}
```

4.5.3 The Function `spotLoop`

(S-5) *Calling the `spotLoop` function.* After the initial phase is finished, the function `spotLoop` is called, which manages the main loop. It is implemented as a stand-alone function, because it can be called separately, e.g., to continue interrupted experiments. With this mechanism, `spot` provides a convenient way for continuing experiments on different computers or extending existing experiments, e.g., if the results are inconclusive or a pre-experimental study should be performed first.

Example: Continue existing experiments

The studies in Sects. 5.8.1, 5.8.2, and 5.8.3 start with a relatively small pre-experimental design. Results from the pre-experimental tests are combined with results from the full experiment.

(S-6) *Consistency Check and Initialization.* Because the `spotLoop` can be used to continue an interrupted `spot` run, it performs a consistency check before the main loop is started.

(S-7) *Imputation.* The function defined by the argument `control$yImputation$handleNAsMethod` is called to handle NA s, Inf s, etc. This is necessary here, because `spotLoop` can be used as an entry point to continue an interrupted `spot` optimization run. How to continue existing `spot` runs is explained in the `spotLoop` documentation.

(S-8) *Counter and Log Data.* Furthermore, counters and logging variables are initialized. The matrix `yBestVec` stores the best function value found so far. It is initialized with the minimum value of the objective function on the initial design. Note, `ySurr`, which keeps track of the objective function values on the surrogate S, has `NA` s, because no surrogate was built so far:

Program Code: Steps (S-5) to (S-8)

```
## (S-5) Enter spotLoop:

## (S-6) Initial check:
initialInputCheck(x, fun, lower, upper, control, inSpotLoop = TRUE)
dimension <- length(lower)
con <- spotControl(dimension)
con[names(control)] <- control
control <- con
rm(con)
control <- spotFillControlList(control, lower, upper)

## (S-7) Imputation:
if (!is.null(control$yImputation$handleNAsMethod)) {
  y <- imputeY(
    x = x,
    y = y,
    control = control
  )
}

## (S-8) Counter and logs:
count <- nrow(y)
modelFit <- NA
ybestVec <- rep(min(y[, 1]), count)
ySurr <- matrix(NA, nrow = 1, ncol = count)
```

4.5.4 Entering the Main Loop

(S-9) *Termination Criteria, Conditions.* The main loop is entered as follows:

```
while ((count < control$funEvals) &
         (difftime(Sys.time(), control$time$startTime, units = 'mins')
          < control$time$maxTime))
```

Two termination criteria are implemented:

a. the number of objective function evaluations must be smaller than
   funEvals and
b. the time must be smaller than maxTime.

(S-10) *Subset Selection for the Surrogate.* Surrogates can be built with the full or a
       reduced set of available x- and y-values. A subset selection method, which is
       defined via control$subsetSelect, can be used before the surrogate
       $S$is built. If subsetSelect is set to selectAll, which is the default, all
       points are used. Fitting the surrogate $S$with a subset of the available points
       only appears to be counterintuitively, but can be reasonable, e.g., if the sample
       points are too close to each other or if the problem changes dynamically.

(S-11) *Fitting the Surrogate.* SPOT can use arbitrary regression models as surrogates,
       e.g., RF or GP models (Kriging).

The arguments x and y are mandatory for the function model. The model
function must return a fit object that provides a predict method. A Gaussian
process model, which performs well in many situations and can work well with
discrete and continuous hyperparameters, is SPOT's default model. Random forest
is less suited as a surrogate for continuous parameters, as it has to approximate
said parameters in a step-wise constant manner. The function control$model
is applied to the x- and y-matrices. A default model is fitted to the data with the
function buildKriging.

Program Code: Steps (S-9) to (S-11)

Steps (S-9) to (S-11) are implemented as follows:

```
## (S-9) Termination (while loop):

## (S-10) Subsect select:
selectRes <- control$subsetSelect(
  x = x,
  y = y[, 1, drop = FALSE],
  control = control$subsetControl
)

## (S-11) Surrogate fit:
modelFit <- control$model(
  x = selectRes$x,
  y = selectRes$y,
  control = control$modelControl
)
```

Table 4.5 Surrogates in `spot` require two arguments, `x` and `y`. The return values of the `build*` functions are shown below

Return Value	Type	Description
x	matrix	x values
y	matrix	y values
fit	object	Fitted model
pNames	character	Names of the independent variables
yName	character	Name of the dependent variable
class	character	Name of the model class

Background: Surrogates

There is a naming convention for surrogates in `spot`: functions names should start with the prefix "`build`". Surrogates in `spot` use the same interface. They accept the arguments `x`, `y`, which must be matrices, and the list `control`. They fit a model, e.g., `buildLM` uses the `lm`, which provides a method `predict`. Each model returns an object of the corresponding model class, here: `"spotLinearModel"`, with a `predict` method. The return value is implemented as a list with the entries from Table 4.5.

Note, `buildLM` is a very simple model. SPOT's workhorse is a Kriging model, that is fitted via Maximum Likelihood Estimation (MLE). `buildKriging` is explained in Sect. 4.6.5.

(S-12) *Objective Function on the Surrogate (Predict).* After building the surrogate, the `modelFit` (surrogate model) is available. It is used to define the function `funSurrogate`, which works as an objective function on the surrogate S: `funSurrogate` does not evaluate solutions on the original function f, but on the surrogate S. Thus, `spot` searches for the hyperparameter configuration that is predicted to result in the best possible model quality. Therefore, an objective function is generated based on the `modelFit` via `predict`.

Program Code: Step (S-12)

Step (S-12) is implemented as follows:

```
## (S-12) Surrogate optimization function:
funSurrogate <- evaluateModel(
  modelFit,
  control$infillCriterion,
  control$verbosity
)
```

Background: Surrogate and Infill Criteria

The function `evaluateModel` generates an objective function that predicts function values on the surrogate. Some surrogate optimization procedures do not use the function values from the surrogate S—they use an infill criterion instead.

Definition 4.2 (*Infill Criterion, Acquisition Function*) Infill criteria are methods that guide the exploration of the surrogate. They combine information from the predicted mean and the predicted variance generated by the GP model. In BO, the term "acquisition function" is used for functions that implement infill criteria.

For example, the function `buildKriging` provides three return values that can be used to generate elementary infill criteria. These return values are specified via the argument `target`, which is a vector of strings. Each string specifies a value to be predicted, e.g., `"y"` for mean, `"s"` for standard deviation, and `"ei"` for expected improvement. In addition to these elementary values, `spot` provides the function `infillCriterion` to specify user-defined criteria. The function `evaluateModel` that manages the infill criteria in `spot` is shown below.

```
evaluateModel <-
  function(object,
           infillCriterion = NULL) {
    evalModelFun <- function(x) {
      res <- predict(object = object, newdata = x)[object$target]
      return(res)
    }
    if (is.null(infillCriterion)) {
      return(function(x) {
        res <- evalModelFun(x)
        return(res)
      })
    } else {
      return(function(x) {
        return(infillCriterion(evalModelFun(x), object))
      })
    }
  }
```

Example: Expected Improvement

EI is a popular infill criterion, which was defined in Eq. (4.10). It is calculated as shown in Eq. (4.11) and can be called from `evaluateModel` via `modelControl = list(target = c("ei")`. The following code shows an EI implementation that returns a vector with the negative logarithm of the expected improvement values, $-\log_{10}(\text{EI})$. The function `expectedImprovement`

is called, if the argument `"ei"` is selected as a `target`, e.g., `spot`
`(,fun,l,u,control=list(modelControl=list(target="ei")))`.

```
expectedImprovement <- function(mean, sd, min) {
  EITermOne = (min - mean) * pnorm((min - mean) / sd)
  EITermTwo = sd * (1 / sqrt(2 * pi))
                  * exp(-(1 / 2) * ((min - mean) ^ 2 / (sd ^ 2)))
  - log10(EITermOne + EITermTwo + (.Machine$double.xmin))
}
```

(S-13) *Multiple Starting Points.* If the current best point is feasible, it is used as
a starting point for the search on the surrogate S. Because the surrogate
can be multi-modal, multiple starting points are recommended. The function `getMultiStartPoints` implements a multi-start mechanism. `spot`
provides the function `getMultiStartPoints`.

In addition to the current best point further starting points can be used. Their
amount can be specified by the value of `multiStart`. If `multiStart` >
1, then additional starting points will be used. The `design` function, which
was used for generating the initial design in Sect. 4.5.2, will be used here to
generate additional points.

(S-14) *Optimization on the Surrogate.* The search on the surrogate Scan be performed next. The simplest objective function is `optimLHD`, which selects
the point with the smallest function value from a relatively large set of
LHD points. Other objective functions are available, e.g., `optimLBFGS`
or `optimDE`. To find the next candidate solution, the predicted value of
the surrogate is optimized via Differential Evolution (Storn and Price 1997).
Other global optimization algorithms can be used as well. Even RS would be
a feasible strategy.

▷ Mandatory Parameters

Optimization functions must use the same interface as `spot`, i.e., `function(x,`
`fun,lower,upper,control=list(),...)`. The arguments `fun`, `lower`,
and `upper` are mandatory for optimization functions. This is similar to the interface
of R's general-purpose optimization function `optim`.

As described in Sect. 4.5.4, the optimization on the surrogate Scan be performed
with or without pre-defined starting points. We describe a search without starting
points first.

(S-14a) *Search Without Starting Points.* If no starting points for the search are provided, the optimizer, which is specified via `control$optimizer`, is called.

The result from this optimization is stored in the list `optimResSurr`. The optimal value from the search on the surrogate is `optimResSurr$xbest`, the corresponding y-value is `optimResSurr$ybest`. Alternatively, the search on the surrogate can be performed with starting points.

(S-14b) *Search With Starting Points.* If starting points are used for the optimization on the surrogate, these are passed via `x = x0` to the optimizer. Several starting points result in several `optimResSurr$xbest` and `optimResSurr$ybest` values from which the best, i.e., the point with the smallest y-value, is selected.

For example, if `multiStart = 2` is selected, the current best and one random point will be used.

The optimization on the surrogate S is performed separately for each starting point and the matrix `xnew` is computed.

`xnew` is determined based on the multi-start results.

(S-15) *Compile Results from the Search on the Surrogate.* The function value of `xnew` (from (S-14a) or (S-14b)) is saved as `ySurrNew`. Note, this function values can be modified using `control$modelControl$target`, e.g., `"y"`, `"s2`, or `"ei"`, i.e., the optimization on the surrogate can be based on the predicted new value `"y"`, a combination of `"y"` and the variance or the EI `"ei"`.

(S-16) *Noise, Repeats, and Consistency Checks for New Points.* After the new solution candidate `xnew` and its associated function value on the surrogate `ySurrNew` have been determined, `spot` checks for duplicates and determines the number of replicates. This step treats noisy and deterministic objective functions in a different way.

If `control$noise == TRUE`, then replicates are allowed, i.e., a single solution *x* can be evaluated several times. If `control$noise == FALSE`, then every solution is evaluated only once.

Program Code: Steps (S-13) to (S-16)

Steps (S-13) to (S-16) are implemented as follows:

```
## (S-13) Random starting points for optimization on the surrogate
x0 <- getMultiStartPoints(x, y, control)
resSurr <- matrix(NA, nrow = nrow(x0), ncol = ncol(x0) + 1)

## (S-14b) Search on the surrogate with starting point/s x0:
for (i in 1:nrow(x0)) {
  optimResSurr <- control$optimizer(
    x = x0[i, , drop = FALSE],
    funSurrogate,
    lower,
```

```
   upper,
   control$optimizerControl
 )
 resSurr[i, ] <- c(optimResSurr$xbest, optimResSurr$ybest)
}

## (S-15) Compile surrogate results:
m <- which.min(resSurr[, ncol(x) + 1])
## Determine xnew based on multi start results
xnew <- resSurr[m, 1:ncol(x), drop = FALSE]
## value on the surrogate (can be "y", "s2, "ei", "negLog10ei" etc.)
ySurrNew <- resSurr[m, ncol(x) + 1]


## (S-16) Duplicate handling:
xnew <- duplicateAndReplicateHandling(xnew, x, lower, upper, control)
# Repair non-numeric results
xnew <- repairNonNumeric(xnew, control$types)
```

Background: Duplicates and Replicates

The function duplicateAndReplicateHandling checks whether the new
solution xnew has been evaluated before. In this case, it is taken as it is and no additional evaluations are performed. If xnew was not evaluated before, it will be evaluated. The number of evaluations is defined via control$replicates. Duplicate
and replicate handling in spot depends on the setting of the parameter noise. If
the value is TRUE then a test whether xnew is new or has been evaluated before is
performed. If xnew is new (was not evaluated before), then it should be evaluated
replicates times. Assume, control$replicates < -3, i.e., three initial
replicates are required and xnew was not evaluated before. Then two additional evaluations should be done, i.e., xtmp contains two entries which are combined with
one already existing entry in xnew.

```
control$replicates <- 3
xtmp <- NULL
for (i in 1:nrow(xnew)) {
  if (!any(apply(x, 1, identical, xnew[i, ]))) {
    xtmp <- rbind(xtmp, xnew[rep(i, control$replicates - 1), ])
  }
}
xnew <- rbind(xnew, xtmp)
xnew
  ##                 [,1]            [,2]
  ## [1,] -0.01292055 -0.02666901
  ## [2,] -0.01292055 -0.02666901
  ## [3,] -0.01292055 -0.02666901
```

If the parameter `noise` has the value FALSE, two cases have to be distinguished. First, if `xnew` was not evaluated before, then it should be evaluated once (and not `replicates` times), because additional evaluation is useless. They would deterministically generate the same result.

```
for (i in 1:nrow(xnew)) {
  if (any(apply(x, 1, identical, xnew[i, ]))) {
    warning("Duplicate is replaced by random solution.")
    control$designControl$replicates <- 1
    control$designControl$size <- 1
    xnew[i, ] <-
      designUniformRandom(, lower, upper, control$designControl)
  }
}
xnew
##                  [,1]           [,2]
## [1,] -0.01292055 -0.02666901
## [2,] -0.01292055 -0.02666901
## [3,] -0.01292055 -0.02666901
```

Second, if `xnew` was evaluated before, then a warning is issued and a randomly generated solution for each entry in `xnew` will be used.

```
# xnew has two already known solutions:
xnew <- x[1:2, ]
for (i in 1:nrow(xnew)) {
  if (any(apply(x, 1, identical, xnew[i, ]))) {
    warning("Duplicate is replaced by random solution.")
    control$designControl$replicates <- 1
    control$designControl$size <- 1
    xnew[i, ] <-
      designUniformRandom(, lower, upper, control$designControl)
  }
}
xnew
##                 [,1]          [,2]
## [1,] 0.6704420 0.1762307
## [2,] 0.5094811 0.3023359
```

A type check is performed, i.e., all non-numeric values produced by the optimizer are rounded.

(S-17) *OCBA for Known Points.* OCBA is called next if OCBA and `noise` are both set to TRUE: the function `repeatsOCBA` returns a vector that specifies how often each known solution should be re-evaluated (or replicated). This function can spend a budget of `control$OCBABudget` additional evaluations. The solutions proposed by `repeatsOCBA` are added to the set of new *x* candidates `xnew`. Because OCBA calculates an estimate of the variance, it is based on evaluated solutions and their function values, i.e., x and y values respectively.

Program Code: Step (S-17)

Step (S-17) is implemented as follows:

```
## (S-17) OCBA:
if (control$noise &
  control$OCBA) {
  xnew <- rbind(xnew, repeatsOCBA(x, y[, 1, drop = FALSE], control$OCBABudget))
}
```

Background: Optimal Computational Budget Allocation

OCBA is a very efficient solution to solve the "general ranking and selection problem" if the objective function is noisy (Chen 2010; Bartz-Beielstein et al. 2011). It allocates function evaluations in an uneven manner to identify the best solutions and to reduce the total optimization costs.

Theorem 4.1 *Given a total number of optimization samples N to be allocated to k competing solutions whose performance is depicted by random variables with means $\bar{y}_i$ (i = 1, 2, ..., k), and finite variances σ_i^2, respectively, as N → ∞, the Approximate Probability of Correct Selection (APCS) can be asymptotically maximized when*

$$\frac{N_i}{N_j} = \left(\frac{\sigma_i/\delta_{b,i}}{\sigma_j/\delta_{b,j}} \right)^2, i, j \in \{1, 2, \ldots, k\}, and\ i \neq j \neq b, \tag{4.1}$$

$$N_b = \sigma_b \sqrt{\sum_{i=1, i \neq b}^{k} \frac{N_i^2}{\sigma_i^2}}, \tag{4.2}$$

where N_i is the number of replications allocated to solution i, $\delta_{b,i} = \bar{y}_b - \bar{y}_i$, and $\bar{y}_b \leq \min_{i \neq b} \bar{y}_i$ (Chen 2010).

(S-18) *Evaluating New Solutions.* To avoid exceeding the available budget of objective function evaluations, which is specified via control$funEvals, a check is performed. Solution candidates are passed to the function objectiveFunctionEvaluation, which calculates the associated objective function values ynew on the function fun.

(S-19) *Imputation.* Because the evaluation of solution candidates might result in infinite Inf or Not-a-Number NaN ynew values, the function imputeY, which handled non-numeric values, is called.

(S-20) *Update Counter and Log Data.* Next, counters `count` and `ySurr`, information about the function values on the surrogate S, are updated.
Calculation of the progress and preparation of progress plots conclude the
main loop. The last step of the main loop compiles the list `return`, which
is returned to the `spot` function.

(S-21) *Reporting after the While-Loop.* After the while loop is finished, results are
compiled. Some objective functions return several values (Multi Objective
Optimization (MOO)). The corresponding values are stored as `logInfo`,
because the default `spot` function uses only one objective function value.
This mechanism enables `spot` handling MOO problems. The values of the
transformed parameters are stored as `xt`. Important for noisy optimization
is the following feature: OCBA can be used for the selection of the best
value. The function `ocbaRanking` computes the best `x` and `y` values,
`xBestOcba` and `yBestOcba`, respectively. `yBestOcba` is the mean value
of the corresponding x-parameter setting `xBestOcba`.

Program Code: Steps (S-18) to (S-22)

Steps (S-18) to (S-22) are implemented as follows:

```
## (S-18) Evaluate xnew:
ynew <- tryCatch(
  expr = {
    objectiveFunctionEvaluation(
      x = x,
      xnew = xnew,
      fun = fun,
      control = control
    )
  },
  error = function(e) {
    if (!is.null(control$yImputation$handleNAsMethod)) {
      n <- nrow(xnew)
      m <- ncol(y)
      return(matrix(rep(NA, m * n), nrow = n))
    }
  }
)
## (S-19) Impute:
colnames(xnew) <- colnames(x)
x <- rbind(x, xnew)
y <- rbind(y, ynew)
if (!is.null(control$yImputation$handleNAsMethod)) {
  y <- imputeY(
    x = x,
    y = y,
    control = control
  )
}
## (S-20) Update counter, logs, etc.:
```

```
ySurr <- c(ySurr, ySurrNew)
count <- count + nrow(ynew)
indexBest <- which.min(y[, 1, drop = FALSE])
ybestVec <- c(ybestVec, y[indexBest, 1, drop = FALSE])
## END while loop
## (S-21) Reporting after while loop in spotLoop
if (ncol(y) > 1) {
  logInfo <- y[, -1, drop = FALSE]
} else {
  logInfo <- NA
}
if (length(control$transformFun) > 0) {
  xt <- transformX(xNat = x, fn = control$transformFun)
} else {
  xt <- NA
}
# (S-22) OCBA-best selection:
if (control$noise & control$OCBA) {
  ocbaRes <- ocbaRanking(
    x = x,
    y = y,
    fun = fun,
    control = control
  )
  control$xBestOcba <- ocbaRes[1, 1:(ncol(ocbaRes) - 1)]
  control$yBestOcba <- ocbaRes[1, ncol(ocbaRes)]
}
# Compile results in spotLoop
result <- list(
  xbest = x[indexBest, , drop = FALSE],
  ybest = y[indexBest, 1, drop = FALSE],
  xBestOcba = matrix(control$xBestOcba, ncol = length(lower)),
  yBestOcba = matrix(control$yBestOcba, ncol = length(lower)),
  x = x,
  xt = xt,
  y = y[, 1, drop = FALSE],
  logInfo = logInfo,
  count = count,
  msg = "budget exhausted",
  modelFit = modelFit,
  ybestVec = ybestVec,
  ySurr = ySurr
)
## END spotLoop()
```

The function spotLoop ends here and the final steps of the main function spot, which are summarized in the following section, are executed.

Table 4.6 `spot`: return parameters

Parameter	Value, type	Description
xbest	matrix	Best x values
ybest	matrix	Best y values
xBestOcba	matrix	Best x values
yBestOcba	matrix	Best y values
x	matrix	x values
xt	matrix	Transformed x values
y	matrix	y values
logInfo	matrix	Additional y information, also multi-objective values
count	integer	Number of function evaluations
msg	character	Information about the optimization
modelFit		
yBestVec	matrix	History of best y values
ySurr	matrix	y values on the surrogate
control	list	Control parameters

4.5.5 Final Steps

To exploit the region of the best solution from the surrogate, S, which was determined during the SMBO in the main loop with `spotLoop`, SPOT allows a local optimization step. If `control$directOptControl$funEvals` is larger than zero, this optimization is started. If the best solution from the surrogate, `xbest`, satisfies the inequality constraints, it is used as a starting point for the local optimization with the local optimizer `control$directOpt`. For example, `directOpt = optimNLOPTR` or `directOpt = optimLBFGSB`, can be used.

Results from the direct optimization will be appended to the matrices of the x and y values based on SMBO. SPOT returns the gathered information in a list (Table 4.6). Because SPOT focuses on reliability and reproducibility, it is not the speediest algorithm.

4.6 Kriging

Basic elements of the Kriging-based surrogate optimization such as interpolation, expected improvement, and regression are presented. The presentation follows the approach described in Forrester et al. (2008a).

4.6.1 The Kriging Model

Consider sample data $\mathbf{X}$ and $\mathbf{y}$ from n locations that are available in matrix form: $\mathbf{X}$ is a $(n \times k)$ matrix, where k denotes the problem dimension and $\mathbf{y}$ is a $(n \times 1)$ vector. The observed responses $\mathbf{y}$ are considered as if they are from a stochastic process, which will be denoted as

$$\begin{pmatrix} \mathbf{Y}(\mathbf{x}^{(1)}) \\ \vdots \\ \mathbf{Y}(\mathbf{x}^{(n)}) \end{pmatrix}.$$

The set of random vectors (also referred to as a "random field") has a mean of $\mathbf{1}\mu$, which is a $(n \times 1)$ vector. The random vectors are correlated with each other using the basis function expression

$$\text{Cor}\left(\mathbf{Y}(\mathbf{x}^{(i)}), \mathbf{Y}(\mathbf{x}^{(l)})\right) = \exp\left\{ -\sum_{j=1}^{k} \theta_j |x_j^{(i)} - x_j^{(l)}|^{p_j} \right\}.$$

The $(n \times n)$ correlation matrix of the observed sample data is

$$\mathbf{\Psi} = \begin{pmatrix} \text{Cor}\left(\mathbf{Y}(\mathbf{x}^{(i)}), \mathbf{Y}(\mathbf{x}^{(l)})\right) & \cdots & \text{Cor}\left(\mathbf{Y}(\mathbf{x}^{(i)}), \mathbf{Y}(\mathbf{x}^{(l)})\right) \\ \vdots & \vdots & \vdots \\ \text{Cor}\left(\mathbf{Y}(\mathbf{x}^{(i)}), \mathbf{Y}(\mathbf{x}^{(l)})\right) & \cdots & \text{Cor}\left(\mathbf{Y}(\mathbf{x}^{(i)}), \mathbf{Y}(\mathbf{x}^{(l)})\right) \end{pmatrix}. \tag{4.3}$$

Note: correlations depend on the absolute distances between sample points $|x_j^{(n)} - x_j^{(n)}|$ and the parameters p_j and θ_j.

To estimate the values of $\boldsymbol{\theta}$ and $\mathbf{p}$, they are chosen to maximize the likelihood of $\mathbf{y}$, which can be expressed as

$$L\left(\mathbf{Y}(\mathbf{x}^{(1)}), \ldots, \mathbf{Y}(\mathbf{x}^{(n)})|\mu, \sigma\right) = \frac{1}{(2\pi\sigma)^{n/2}} \exp\left\{ \frac{-\sum_{j=1}^{n} \left(\mathbf{Y}^{(j)} - \mu\right)^2}{2\sigma^2} \right\},$$

which can be expressed in terms of the sample data

$$L\left(\mathbf{Y}(\mathbf{x}^{(1)}), \ldots, \mathbf{Y}(\mathbf{x}^{(n)})|\mu, \sigma\right) = \frac{1}{(2\pi\sigma)^{n/2}|\mathbf{\Psi}|^{1/2}} \exp\left\{ \frac{-(\mathbf{y} - \mathbf{1}\mu)^T \mathbf{\Psi}^{-1}(\mathbf{y} - \mathbf{1}\mu)}{2\sigma^2} \right\},$$

and formulated as the log-likelihood:

$$\ln(L) = -\frac{n}{2}\ln(2\pi\sigma) - \frac{1}{2}\ln|\mathbf{\Psi}| \frac{-(\mathbf{y} - \mathbf{1}\mu)^T \mathbf{\Psi}^{-1}(\mathbf{y} - \mathbf{1}\mu)}{2\sigma^2}. \tag{4.4}$$

Optimization of the log-likelihood by taking derivatives with respect to μ and σ results in

$$\hat{\mu} = \frac{\mathbf{1}^T \mathbf{\Psi}^{-1} \mathbf{y}^T}{\mathbf{1}^T \mathbf{\Psi}^{-1} \mathbf{1}^T} \tag{4.5}$$

and

$$\hat{\sigma} = \frac{(\mathbf{y} - \mathbf{1}\mu)^T \mathbf{\Psi}^{-1} (\mathbf{y} - \mathbf{1}\mu)}{n}. \tag{4.6}$$

Substituting (4.5) and (4.6) into (4.4) leads to the concentrated log-likelihood:

$$\ln(L) = -\frac{n}{2} \ln(\hat{\sigma}) - \frac{1}{2} \ln |\mathbf{\Psi}|. \tag{4.7}$$

Note: To maximize $\ln(L)$, optimal values of θ and $\mathbf{p}$ are determined numerically, because (4.7) is not differentiable.

4.6.2 Kriging Prediction

For a new prediction $\hat{y}$ at $\mathbf{x}$, the value of $\hat{y}$ is chosen so that it maximizes the likelihood of the sample data $\mathbf{X}$ and the prediction, given the correlation parameter θ and $\mathbf{p}$. The observed data $\mathbf{y}$ is augmented with the new prediction $\hat{y}$ which results in the augmented vector $\tilde{\mathbf{y}} = (\mathbf{y}^T, \hat{y})^T$. A vector of correlations between the observed data and the new prediction is defined as

$$\psi = \begin{pmatrix} \text{Cor}\left(\mathbf{Y}(\mathbf{x}^{(1)}), \mathbf{Y}(\mathbf{x})\right) \\ \vdots \\ \text{Cor}\left(\mathbf{Y}(\mathbf{x}^{(n)}), \mathbf{Y}(\mathbf{x})\right) \end{pmatrix} = \begin{pmatrix} \psi^{(1)} \\ \vdots \\ \psi^{(n)} \end{pmatrix}.$$

The augmented correlation matrix is constructed as

$$\tilde{\mathbf{\Psi}} = \begin{pmatrix} \mathbf{\Psi} & \psi \\ \psi^T & 1 \end{pmatrix}.$$

Similar to (4.4), the log-likelihood of the augmented data is

$$\ln(L) = -\frac{n}{2} \ln(2\pi) - \frac{n}{2} \ln(\hat{\sigma}^2) - \frac{1}{2} \ln |\hat{\mathbf{\Psi}}| - \frac{(\tilde{\mathbf{y}} - \mathbf{1}\hat{\mu})^T \tilde{\mathbf{\Psi}}^{-1} (\tilde{\mathbf{y}} - \mathbf{1}\hat{\mu})}{2\hat{\sigma}^2}. \tag{4.8}$$

The MLE for $\hat{y}$ can be calculated as

$$\hat{y}(\mathbf{x}) = \hat{\mu} + \psi^T \tilde{\mathbf{\Psi}}^{-1} (\mathbf{y} - \mathbf{1}\hat{\mu}). \tag{4.9}$$

Equation 4.9 reveals two important properties of the Kriging predictor.

- The basis function impacts the vector $\boldsymbol{\psi}$, which contains the n correlations between the new point $\mathbf{x}$ and the observed locations. Values from the n basis functions are added to a mean base term μ with weightings $\mathbf{w} = \tilde{\boldsymbol{\Psi}}^{(-1)}(\mathbf{y} - \mathbf{1}\hat{\mu})$.
- The predictions interpolate the sample data. When calculating the prediction at the ith sample point, $\mathbf{x}^{(i)}$, the ith column of $\boldsymbol{\Psi}^{-1}$ is $\boldsymbol{\psi}$, and $\boldsymbol{\psi}\boldsymbol{\Psi}^{-1}$ is the ith unit vector. Hence, $\hat{y}(\mathbf{x}^{(i)}) = y^{(i)}$.

4.6.3 Expected Improvement

The EI is a criterion for error-based exploration, which uses the MSE of the Kriging prediction. The MSE is calculated as

$$s^2(\mathbf{x}) = \sigma^2 \left(1 - \boldsymbol{\psi}^T \boldsymbol{\Psi}^{-1} \boldsymbol{\psi} + \frac{(1 - \mathbf{1}^T \boldsymbol{\Psi}^{-1} \boldsymbol{\psi})^2}{\mathbf{1}^T \boldsymbol{\Psi}^{-1} \mathbf{1}} \right).$$

Here, $s^2(\mathbf{x}) = 0$ at sample points, and the last term is omitted in Bayesian settings.

Since the EI extends the Probability of Improvement (PI), it will be described first. Let $y_{\min}$ denote the best-observed value so far and consider $\hat{y}(\mathbf{x})$ as the realization of a random variable. Then, the probability of an improvement $I = y_{\min} - Y(\mathbf{x})$ can be calculated as

$$P(I(\mathbf{x})) = \frac{1}{2} \left\{ 1 + \mathrm{erf}\left(\frac{y_{\min} - \hat{y}(\mathbf{x})}{\hat{s}\sqrt{2}} \right) \right\}.$$

The EI does not calculate the probability that there will be some improvement, it calculates the amount of expected improvement. The rationale of using this expectation is that we are less interested in highly probable improvement if the magnitude of that improvement is very small. The EI is defined as follows.

Definition 4.3 (*Expected Improvement*)

$$E(I(\mathbf{x})) = \begin{cases} (y_{\min} - \hat{y}(\mathbf{x}))\Phi\left(\frac{y_{\min} - \hat{y}(\mathbf{x})}{\hat{s}(\mathbf{x})} \right) + \hat{s}\phi\left(\frac{y_{\min} - \hat{y}(\mathbf{x})}{\hat{s}(\mathbf{x})} \right) & \text{if } \hat{s} > 0 \\ 0 & \text{if } \hat{s} = 0 \end{cases}, \qquad (4.10)$$

where $\Phi(.)$ and $\phi(.)$ are the Cumulative Distribution Function (CDF) and Probability Distribution Function (PDF), respectively.

The EI is evaluated as

$$E(I(\mathbf{x})) = (y_{\min} - \hat{y}(\mathbf{x})) \frac{1}{2} \left\{ 1 + \mathrm{erf}\left(\frac{y_{\min} - \hat{Y}(\mathbf{x})}{\hat{s}\sqrt{2}} \right) \right\} + \hat{s} \frac{1}{\sqrt{2\pi}} \exp\left\{ \frac{-(y_{\min} - \hat{y}(\mathbf{x}))^2}{2\hat{s}^2} \right\}.$$
$$(4.11)$$

4.6.4 Infill Criteria with Noisy Data

The EI infill criterion was formulated under the assumption that the true underlying function is deterministic, smooth, and continuous. In deterministic settings, the Kriging predictor should interpolate the data. Noise can complicate the modeling process: predictions can become erratic, because there is a high MSE in regions far away from observed data. Therefore, the interpolation property should be dropped to filter noise. A regression constant, λ, is added to the diagonal of $\mathbf{\Psi}$ and $\mathbf{\Psi} + \lambda\mathbf{I}$ is used. Then, $\mathbf{\Psi} + \lambda\mathbf{I}$ does not contain ψ as a column and the data is not interpolated. The same method of derivation as in interpolating Kriging (Eq. 4.9) can be used for regression Kriging. The regression Kriging prediction is given by

$$\hat{y}_r(\mathbf{x}) = \hat{\mu}_r + \boldsymbol{\psi}^T(\mathbf{\Psi} + \lambda\mathbf{I})^{-1}(\mathbf{y} - \mathbf{1}\hat{\mu}_r),$$

where

$$\hat{\mu}_r = \frac{\mathbf{1}^T(\mathbf{\Psi} + \lambda\mathbf{I})^{-1}\mathbf{y}}{\mathbf{1}^T(\mathbf{\Psi} + \lambda\mathbf{I})^{-1}\mathbf{1}}.$$

Including the regression constant λ the following equation allows the calculation of an estimate of the error in the Kriging regression model for noisy data:

$$\hat{s}_r^2(\mathbf{x}) = \hat{\sigma}_r^2\left\{1 + \lambda - \boldsymbol{\psi}^T(\mathbf{\Psi} + \lambda\mathbf{I})^{-1}\boldsymbol{\psi} + \frac{(1 - \mathbf{1}^T(\mathbf{\Psi} + \lambda\mathbf{I})^{-1}\boldsymbol{\psi})^2}{\mathbf{1}^T(\mathbf{\Psi} + \lambda\mathbf{I})^{-1}\mathbf{1}}\right\}, \qquad (4.12)$$

where

$$\hat{\sigma}_r^2 = \frac{(\mathbf{y} - \mathbf{1}\hat{\mu}_r)^T(\mathbf{\Psi} + \lambda\mathbf{I})^{-1}(\mathbf{y} - \mathbf{1}\hat{\mu}_r)}{n}.$$

Note: Eq. (4.12) includes the error associated with noise in the data. There is nonzero error in all areas which leads to nonzero EI in all areas. As a consequence, resampling can occur. Resampling can be useful if replicates result in different outcomes. Although the possibility of resampling can destroy the convergence to the global optimum, resampling can be a wanted feature in optimization with noisy data. In a deterministic setting, resampling is an unwanted feature, because new evaluations of the same point do not provide additional information and can stall the optimization process.

Re-interpolation can be used to eliminate the errors due to noise in the data from the model. Re-interpolation bases the estimated error on an interpolation of points predicted by the regression model at the sample locations. It proceeds as follows: calculate values for the Kriging regression at the sample locations using

$$\hat{\mathbf{y}}_r = \mathbf{1}\hat{\mu} + \mathbf{\Psi}(\mathbf{\Psi} + \lambda\mathbf{I})^{-1}(\mathbf{y} - \mathbf{1}\hat{\mu}).$$

This vector can be substituted into Eq. (4.9), which is substituted into (4.6). This results in

$$\hat{\sigma}_{ri}^2 = \frac{(\mathbf{y} - \mathbf{1}\hat{\mu})^T (\mathbf{\Psi} + \lambda \mathbf{I})^{-1} \mathbf{\Psi} (\mathbf{\Psi} + \lambda \mathbf{I})^{-1} (\mathbf{y} - \mathbf{1}\hat{\mu})}{n}.$$

Using the interpolating Kriging error estimate (4.12), the re-interpolation error estimate reads

$$\hat{s}_{ri}^2(\mathbf{x}) = \hat{\sigma}_{ri}^2 \left\{ 1 - \boldsymbol{\psi}^T \mathbf{\Psi}^{-1} \boldsymbol{\psi} + \frac{(1 - \mathbf{1}^T (\mathbf{\Psi} + \lambda \mathbf{I})^{-1} \boldsymbol{\psi})^2}{\mathbf{1}^T (\mathbf{\Psi} + \lambda \mathbf{I})^{-1} \mathbf{1}} \right\}.$$

4.6.5   spot's Workhorse: Kriging

This section explains the implementation of the function `buildKriging` in SPOT.

(K-1) *Set Parameters.* `buildKriging` uses the parameters shown in Table 4.7. It returns an object of class `kriging`, which is basically a list, with the options and found parameters for the model which has to be passed to the `predict` method of this class.

Program Code: Step (K-1)

```
buildKriging <- function(x, y, control = list()) {
  ## (K-1) Set Parameters
  k <- ncol(x)  # dimension
  n <- nrow(x)  # number of observations
  con <- list(
    thetaLower = 1e-4,
    thetaUpper = 1e2,
    types = rep("numeric", k),
    algTheta = optimDE,
    budgetAlgTheta = 200,
    optimizeP = FALSE,
    useLambda = TRUE,
    lambdaLower = -6,
    lambdaUpper = 0,
    startTheta = NULL,
    reinterpolate = TRUE,
    target = "y"
  )
  fit <- control
  fit$x <- x
  fit$y <- y
  LowerTheta <- rep(1, k) * log10(fit$thetaLower)
  UpperTheta <- rep(1, k) * log10(fit$thetaUpper)
```

Table 4.7 `buildKriging`: besides the design matrix x with corresponding observations y, the function accepts a list with the parameters shown below

Parameter	Value, type	Description
types	character vector	A character vector giving the data type of each variable. All but `factor` will be handled as numeric, `factor` (categorical) variables will be subject to the Hamming distance
thetaLower	1e-4, numerical	Lower boundary for `theta`
thetaUpper	1e2	Upper boundary for `theta`
algTheta	optimDE, function	Algorithm used to find `theta` via MLE
budgetAlgTheta	200, integer	Budget for the algorithm `algTheta`. The value will be multiplied with the length of the model parameter vector to be optimized
optimizeP	FALSE, logical	Specifies whether the exponents (p) should be optimized. Otherwise, they will be set to two
useLambda	TRUE, logical	Whether to use the regularization constant lambda (nugget effect)
lambdaLower	−6, numerical	Lower boundary for `log10(lambda)`
lambdaUpper	0, numerical	Upper boundary for `log10(lambda)`
startTheta	NULL, numerical	Optional start value for theta optimization
reinterpolate	TRUE, logical	Whether re-interpolation should be performed
target	"y", character vector	Values of the prediction. Each element specifies a value to be predicted, e.g., "y" for mean, "s" for standard deviation, "ei" for EI

(K-2) *Normalization.*

The function `normalizeMatrix` is used to normalize the data, i.e., each column of the (n, k)-matrix X has values in the range from zero to one.

Program Code: Step (K-2)

```
## (K-2) Normalize input data
fit$normalizeymin <- 0
fit$normalizeymax <- 1
res <- normalizeMatrix(fit$x, ymin, ymax)
fit$scaledx <- res$y
fit$normalizexmin <- res$xmin
fit$normalizexmax <- res$xmax
```

(K-3) *Correlation Matrix.* Prepare correlation matrix Ψ (Eq. (4.3)) and start points for the optimization. The distance matrix is determined. The i-th row of (k, n^2)-

matrix A contains the distances between the elements of the i-th column (dimension). $A(1, 1)$ is the distance of the first element to the first element in the first dimension, $A(1, 2)$ the distance of the first element to the second element in the first dimension, $A(1, n + 1)$ is the distance of the second element to the first element in the first dimension, and so on.

Program Code: Step (K-3)

```
## (K-3) Prepare distance/correlation matrix
A <- matrix(0, k, n * n)
for (i in 1:k) {
  if (control$types[i] != "factor") {
    A[i, ] <-
      as.numeric(as.matrix(dist(fit$scaledx[, i]))) # euclidean distance
  } else {
    tmp <-
      outer(fit$scaledx[, i], fit$scaledx[, i], "!=") # hamming distance
    class(tmp) <- "numeric"
    A[i, ] <- tmp
  }
}
```

(K-4) *Prepare Starting Points.*

(K-4.1) θ. The starting point for the optimization of θ is determined. If no explicit starting point is specified, then

$$\theta_0 = n/(100k) \tag{4.13}$$

is chosen.

(K-4.2) *p*. The parameter optimizeP determines whether p should be optimized or not. In the latter case, $p = 2$ is set and the matrix A is squared. Otherwise, the starting point for the optimization of p is chosen as $p_0 = 1.9$ and the search interval is set to $[0.01, 2]$.

(K-4.3) *Nugget.* If a nugget effect should be integrated, the starting point for the optimization of λ is set to

$$\lambda_0 = \frac{\lambda_{\text{lower}} + \lambda_{\text{upper}}}{2} \tag{4.14}$$

(K-4.4) *Penalty.* The penalty value is set to

$$\phi = n \times \log(\text{Var}(y)) + 1e4. \tag{4.15}$$

Note: this penalty value should not be a hard constant. The scale of the likelihood, i.e., $n \times \log(SigmaSqr) + LnDetPsi$ at least depends on $\log(\text{Var}(y))$ and the number of samples. Hence, large number of samples may lead to cases where the penalty is lower than the likelihood of most valid parameterizations. A suggested penalty is therefore $\phi = n \times \log(\text{Var}(y)) + 1e4$. Currently, this penalty is set in the `buildKriging` function, when calling `krigingLikelihood`.

Program Code: Step (K-4)

```
## (K-4) Prepare starting points, search bounds and penalty value
## for MLE optimization
## 4.1 theta
x1 <- rep(n / (100 * k), k) # start point for theta
## 4.2 p
LowerTheta <- c(LowerTheta, rep(1, k) * 0.01)
UpperTheta <- c(UpperTheta, rep(1, k) * 2)
x3 <- rep(1, k) * 1.9 # start values for p
x0 <- c(x1, x3)
## 4.3 lambda
# start value for lambda:
x2 <- (fit$lambdaUpper + fit$lambdaLower) / 2
x0 <- c(x0, x2)
# append regression constant lambda (nugget)
LowerTheta <- c(LowerTheta, fit$lambdaLower)
UpperTheta <- c(UpperTheta, fit$lambdaUpper)
x0 <- matrix(x0, 1) # matrix with one row
opts <- list(funEvals = fit$budgetAlgTheta * ncol(x0))
## 4.4 penalty
penval <- n * log(var(y)) + 1e4
```

(K-5) *Objective.* The objective function `fitFun` for the MLE optimizer `algTheta` is defined in this step.

(K-6) *krigingLikelihood.* The function `krigingLikelihood`, see Sect. 4.6.6, is called.

Program Code: Steps (K-5) and (K-6)

```
## (K-5) MLE objective function
fitFun <-
  function(x, fX, fy, optimizeP, useLambda, penval) {
    krigingLikelihood(x, fX, fy, optimizeP, useLambda, penval)$NegLnLike
  }

## (K-6) See krigingLikelihood
```

(K-7) *Performing the Optimization with* $fitFun$. The optimizer is called as follows:

Program Code: Step (K-7)

```
## (K-7)   MLE optimization
res <- fit$algTheta(
  x = x0,
  fun =
    function(x, fX, fy, optimizeP, useLambda, penval) {
      apply(x, 1, fitFun, fX, fy, optimizeP, useLambda, penval)
    },
  lower = LowerTheta,
  upper = UpperTheta,
  control = opts,
  fX = A,
  fy = fit$y,
  optimizeP = fit$optimizeP,
  useLambda = fit$useLambda,
  penval = penval
)
```

(K-8) *Compile Results.* Step Compile return values: The return values from the optimization run, which are stored in the list res, are added to the list fit that specifies the object of the class kriging. The list fit contains the following optimized values: θ^* as Theta, 10^{θ^*} as dmodeltheta, p^*, as P, λ^*, as Lambda and 10^{λ^*}, as dmodellambda.

Program Code: Step (K-8)

```
## (K-8) Compile results from MLE optimization (to fit object)
Params <- res$xbest
nevals <- as.numeric(res$count[[1]])
fit$Theta <- Params[1:k]
fit$dmodeltheta <- 10^Params[1:k]
fit$P <- Params[(k + 1):(2 * k)]
fit$Lambda <- Params[length(Params)]
fit$dmodellambda <- 10^Params[length(Params)]
```

(K-9) *Use Results to Determine Likelihood and Best Parameters.* The function krigingLikelihood is called with these optimized values, θ^*, p^*, and λ^* to determine the values used for the fit of the Kriging model.

Program Code: Step (K-9)

```
## (K-9) Evaluate with optimized parameters
res <-
  krigingLikelihood(
    c(fit$Theta, fit$P, fit$Lambda),
    A,
    fit$y,
    fit$optimizeP,
    fit$useLambda
  )
```

(K-10) *Compile the fit Object.* The return values from this call to krigingLikelihood are added to the fit object.

Program Code: Step (K-10)

```
## (K-10) Add results from MLE evaluation to fit object
fit$yonemu <- res$yonemu
fit$ssq <- as.numeric(res$ssq)
fit$mu <- res$mu
fit$Psi <- res$Psi
fit$Psinv <- res$Psinv
fit$nevals <- nevals
```

```
fit$like <- res$NegLnLike
fit$returnCrossCor <- FALSE
```

(K-11) *Calculate the mean objective function value.* In addition to the results from
the MLE optimization, the mean objective function value of the best x value,
y_{min}, is calculated and stored in the fit list as min. This value is needed
for the EI computation.
Now the fit is available and can be used for predictions. The corresponding
code is shown below.

4.6.6 krigingLikelihood

Step: MLE optimization with $krigingLikelihood$. The objective function
accepts the following parameters: x, a vector, which contains the parameters
log10(theta), log10(lambda), and p, AX, a three-dimensional array, constructed by
$buildKriging$ from the sample locations, Ay, a vector of observations at sample
locations, $optimizeP$, logical, which specifies whether or not to optimize parameter p (exponents) or fix at two, $useLambda$, logical, which specifies whether to use
the nugget, and $penval$, a penalty value which affects the value returned for invalid
correlation matrices or configurations. The function $krigingLikelihood$ performs the following calculations: The θ and λ values are updated:

$$\theta_j = 10^{\theta_0} \quad (j = 1, \ldots, n) \tag{4.16}$$

$$\lambda = 10^{\lambda} \tag{4.17}$$

$$AX[j,] = |(|AX)_j^p \quad (j = 1, \ldots, n) \tag{4.18}$$

(L-1) *Starting Points.*
(L-2) *Correlation Matrix* Ψ. The matrix Ψ can be calculated. If
$useLambda == TRUE$, the nugget effect λ is added.
(L-3) *Cholesky Factorization.* Since $\Psi > 0$, its Cholesky factorization is computed.
(L-4) *Determinant.* The natural log of the determinant of Ψ, $LnDetPsi$ is calcu-
lated, because it is numerically more reliable and also faster than using det
or $determinant$.
(L-5) *Matrix inverse, mean, error, and likelihood.*
Using $chol2inv$, the following values can be calculated: $\ln(L)$ (Eq. (4.7)),
$\hat{\mu}$ (Eq. (4.5)), and $\hat{\sigma}$ (Eq. (4.6)). Together with the matrices Ψ and Ψ^{-1}, and
the vector 1μ, these values are combined into a list, which is returned from
the function $krigingLikelihood$.

The following code illustrates the main components of $krigingLikelihood$.

Program Code: krigingLikelihood Function

```r
krigingLikelihood <-
  function(x,
           AX,
           Ay,
           optimizeP = FALSE,
           useLambda = TRUE,
           penval = 1e8) {
    ## (L-1) Starting Points
    nx <- nrow(AX)
    theta <- 10^x[1:nx]
    if (optimizeP) {
      AX <- abs(AX)^(x[(nx + 1):(2 * nx)])
    }
    lambda <- 0
    if (useLambda) {
      lambda <- 10^x[length(x)]
    }
    n <- dim(Ay)[1]
    ## (L-2) Correlation Matrix Psi
    Psi <- exp(-matrix(colSums(theta * AX), n, n))
    if (useLambda) {
      Psi <- Psi + diag(lambda, n)
    }
    ## (L-3) cholesky decomposition
    cholPsi <- try(chol(Psi), TRUE)
    ## (L-4) Determininant
    LnDetPsi <- 2 * sum(log(abs(diag(cholPsi))))
    ## (L-5.1) Psi Inverted
    Psinv <- try(chol2inv(cholPsi), TRUE)
    psisum <- sum(Psinv)
    ## (L-5.2) Mean
    mu <- sum(Psinv %*% Ay) / psisum
    ## (L-5.3) yoneMu, SigmSqr
    yonemu <- Ay - mu
    SigmaSqr <- (t(yonemu) %*% Psinv %*% yonemu) / n
    ## (L-5.4) Log Likelihood
    NegLnLike <- n * log(SigmaSqr) + LnDetPsi
    ## (L-5.5) Compile Result
    list(
      NegLnLike = NegLnLike,
      Psi = Psi,
      Psinv = Psinv,
      mu = mu,
      yonemu = yonemu,
      ssq = SigmaSqr
    )
  }
```

4.6.7 Predictions

The `buildKriging` function from the R package `spot` provides two Kriging predictors: prediction with and without re-interpolation. Re-interpolation is presented here, because it prevents an incorrect approximation of the error which might cause a poor global convergence. Re-interpolation bases the computation of the estimated error on an interpolation of points predicted by the regression model at the sample locations, see Forrester et al. (2008a).

The function `predictKrigingReinterpolation` requires two arguments: (i) `object`, the Kriging model (settings and parameters) of class `kriging`, and (ii) `newdata`, the design matrix to be predicted.

The function `normalizeMatrix2` is used to normalize the data. It uses information from the normalization performed during the Kriging model building phase, namely `normalizexmin` and `normalizexmax` to ensure the same scaling of the known and new data. Furthermore, the following optimized parameters from the Kriging model are extracted: `scaledx`, `dmodeltheta`, `dmodellambda`, `Psi`, `Psinv`, `mu`, and `yonemu`.

For re-interpolation, the error in the model excluding the error caused by noise is computed. The following modifications are made:

```
PsiB <-
  Psi - diag(lambda, n) + diag(.Machine$double.eps, n)
SigmaSqr <-
  as.numeric(t(yonemu) %*% Psinv %*% PsiB %*% Psinv %*% yonemu) /
    n
Psinv <- try(solve.default(PsiB), TRUE)
if (class(Psinv)[1] == "try-error") {
  Psinv <- ginv(PsiB)
}
```

The MLE for $\hat{y}$ is

$$\hat{y}(x) = \hat{\mu} + \psi^T \Psi^{-1}(y - 1\hat{\mu}). \tag{4.19}$$

This is Eq. (2.40) in Forrester et al. (2008a). It is implemented as follows:

```
psi <- matrix(0, k, n)
for (i in 1:nvar) {
  tmp <- expand.grid(AX[, i], x[, i])
  if (object$types[i] == "factor") {
    tmp <- as.numeric(tmp[, 1] != tmp[, 2])^p[i]
  } else {
    tmp <- abs(tmp[, 1] - tmp[, 2])^p[i]
  }
  psi <- psi + theta[i] * matrix(tmp, k, n, byrow = TRUE)
}

psi <- exp(-psi)


f <- mu + as.numeric(psi %*% (Psinv %*% yonemu))
```

Depending on the setting of the parameter `target`, the values `y` and `s` or `y`, `s`, and `ei` are returned.

```
res <- list(y = f)
if (any(object$target %in% c("s", "ei"))) {
  #
  Psinv <- try(solve.default(PsiB), TRUE)
  if (class(Psinv)[1] == "try-error") {
    Psinv <- ginv(PsiB)
  }
  #
  SSqr <-
    SigmaSqr * (1 - diag(psi %*% (Psinv %*% t(psi))))
  s <- sqrt(abs(SSqr))
  res$s <- s
  if (any(object$target == "ei")) {
    res$ei <- expectedImprovement(f, s, object$min)
  }
}
if (object$returnCrossCor) {
  res$psi <- psi
}
res
```

4.7   Program Code

One complete `spot` run is shown below. To increase readability, only one iteration of the `spotLoop` is performed.

Program Code: spot Run

```
## (S-1) Setup:
fun <- funNoise
lower <- c(-1, -1)
upper <- c(1, 1)
control <- list(
  OCBA = TRUE,
  OCBABudget = 3,
  replicates = 2,
  noise = TRUE,
  multiStart = 2,
  designControl = list(replicates = 2)
)
control <- spotFillControlList(control, lower, upper)

## (S-2) Initial design:
set.seed(control$seedSPOT)
```

```r
x <- control$design(
  x = NULL,
  lower = lower,
  upper = upper,
  control = control$designControl
)
x <- repairNonNumeric(x, control$types)

## (S-3) Eval initial design
y <- objectiveFunctionEvaluation(
  x = NULL,
  xnew = x,
  fun = fun,
  control = control
)

## (S-4) Imputation
if (!is.null(control$yImputation$handleNAsMethod)) {
  y <- imputeY(
    x = x,
    y = y,
    control = control
  )
}

## (S-5) Enter spotLoop:

## (S-6) Initial check:
initialInputCheck(x, fun, lower, upper, control, inSpotLoop = TRUE)
dimension <- length(lower)
con <- spotControl(dimension)
con[names(control)] <- control
control <- con
rm(con)
control <- spotFillControlList(control, lower, upper)

## (S-7) Imputation:
if (!is.null(control$yImputation$handleNAsMethod)) {
  y <- imputeY(
    x = x,
    y = y,
    control = control
  )
}

## (S-8) Counter and logs:
count <- nrow(y)
modelFit <- NA
ybestVec <- rep(min(y[, 1]), count)
ySurr <- matrix(NA, nrow = 1, ncol = count)

## (S-9) Termination (while loop):

## (S-10) Subsect select:
```

```r
selectRes <- control$subsetSelect(
  x = x,
  y = y[, 1, drop = FALSE],
  control = control$subsetControl
)

## (S-11) Surrogate fit:
modelFit <- control$model(
  x = selectRes$x,
  y = selectRes$y,
  control = control$modelControl
)

## (S-12) Surrogate optimization function:
funSurrogate <- evaluateModel(
  modelFit,
  control$infillCriterion,
  control$verbosity
)

## (S-13) Random starting points: surrogate optimization
x0 <- getMultiStartPoints(x, y, control)
resSurr <- matrix(NA, nrow = nrow(x0), ncol = ncol(x0) + 1)

## (S-14b) Surrogate optimization:
for (i in 1:nrow(x0)) {
  optimResSurr <- control$optimizer(
    x = x0[i, , drop = FALSE],
    funSurrogate,
    lower,
    upper,
    control$optimizerControl
  )
  resSurr[i, ] <- c(optimResSurr$xbest, optimResSurr$ybest)
}

## (S-15) Compile surrogate results:
m <- which.min(resSurr[, ncol(x) + 1])
## Determine xnew based on multi start results
xnew <- resSurr[m, 1:ncol(x), drop = FALSE]
## value on the surrogate (can be "y", "s2, "ei", "negLog10ei" etc.)
ySurrNew <- resSurr[m, ncol(x) + 1]


## (S-16) Duplicate handling:
xnew <- duplicateAndReplicateHandling(xnew, x, lower, upper, control)
# Repair non-numeric results
xnew <- repairNonNumeric(xnew, control$types)

## (S-17) OCBA
if (control$noise & control$OCBA) {
  xnew <- rbind(xnew, repeatsOCBA(
    x, y[, 1, drop = FALSE],
    control$OCBABudget
```

```
  ))
}

## (S-18) Evaluate xnew
ynew <- tryCatch(
  expr = {
    objectiveFunctionEvaluation(
      x = x,
      xnew = xnew,
      fun = fun,
      control = control
    )
  },
  error = function(e) {
    message("Error in objectiveFunctionEvaluation()!")
    print(e)
    if (!is.null(control$yImputation$handleNAsMethod)) {
      message("Error will be corrected.")
      n <- nrow(xnew)
      m <- ncol(y)
      return(matrix(rep(NA, m * n), nrow = n))
    }
  }
)

## (S-19) Impute
colnames(xnew) <- colnames(x)
x <- rbind(x, xnew)
y <- rbind(y, ynew)
if (!is.null(control$yImputation$handleNAsMethod)) {
  y <- imputeY(
    x = x,
    y = y,
    control = control
  )
}

## (S-20) Update counter, logs, etc.
ySurr <- c(ySurr, ySurrNew)
count <- count + nrow(ynew)
indexBest <- which.min(y[, 1, drop = FALSE])
ybestVec <- c(ybestVec, y[indexBest, 1, drop = FALSE])

## END while loop

## (S-21) Reporting after while loop in spotLoop
if (ncol(y) > 1) {
  logInfo <- y[, -1, drop = FALSE]
} else {
  logInfo <- NA
}
if (length(control$transformFun) > 0) {
  xt <- transformX(xNat = x, fn = control$transformFun)
} else {
```

```r
  xt <- NA
}
## (S-22) OCBA-based selection of the best
if (control$noise & control$OCBA) {
  ocbaRes <- ocbaRanking(
    x = x,
    y = y,
    fun = fun,
    control = control
  )
  control$xBestOcba <- ocbaRes[1, 1:(ncol(ocbaRes) - 1)]
  control$yBestOcba <- ocbaRes[1, ncol(ocbaRes)]
}
# Compile results in spotLoop
result <- list(
  xbest = x[indexBest, , drop = FALSE],
  ybest = y[indexBest, 1, drop = FALSE],
  xBestOcba = matrix(control$xBestOcba, ncol = length(lower)),
  yBestOcba = matrix(control$yBestOcba, ncol = length(lower)),
  x = x,
  xt = xt,
  y = y[, 1, drop = FALSE],
  logInfo = logInfo,
  count = count,
  msg = "budget exhausted",
  modelFit = modelFit,
  ybestVec = ybestVec,
  ySurr = ySurr
)
## END spotLoop()

if (control$directOptControl$funEvals > 0) {
  ## (S-23) Starting point for direct optimization
  xbest <- result$xbest
  if (!is.null(control$directOptControl$eval_g_ineq) &&
    (
      control$directOptControl$opts$algorithm == "NLOPT_GN_ISRES" &
        control$directOptControl$eval_g_ineq(xbest) < 0
    )) {
    x0 <- NULL
  } else {
    x0 <- xbest
  }

  # Direct optimization on the real fun
  optimResDirect <- control$directOpt(
    x = x0,
    fun = fun,
    lower = lower,
    upper = upper,
    control$directOptControl
  )

  ## (S-24) Update results adding direct
```

```
  if (result$ybest > optimResDirect$ybest) {
    result$xbest <- optimResDirect$xbest
    result$ybest <- optimResDirect$ybest
  }
  result$x <- rbind(result$x, optimResDirect$x)
  result$y <- rbind(result$y, optimResDirect$y)
}
```

The result from one `spotLoop` is saved in the variable `result`.

Chapter 5
Ranking and Result Aggregation

Thomas Bartz-Beielstein, Olaf Mersmann, and Sowmya Chandrasekaran

Abstract This chapter explores different methods to analyze the results of Hyper-parameter Tuning (HPT) experiments. Four different scenarios and two different approaches are presented. On the one hand, rankings and especially consensus rankings are introduced to aggregate the results of many different HPT results. On the other hand, statistical significance analysis and power analysis are used for a detailed analysis of single algorithms and pairwise algorithm comparisons. This chapter discusses issues with sample size determination, power calculations, hypotheses, and wrong conclusions from hypothesis testing. On top of the established methods, we add and explain severity, a frequentist approach that extends the classical concept of p-values. Mayo's concept of severity offers one solution to these issues, and one might achieve even better results by applying severity.

5.1 Comparing Algorithms

Aggregating the results of any kind of hyperparameter tuning or other large-scale modeling experiment poses its own set of challenges. Generally, we can differentiate between four settings (Bartz-Beielstein and Preuss 2011):

Definition 5.1 [Algorithm-Problem Designs]

Single Algorithm Single Problem (SASP): Analyzing the result of a single algorithm or learner on a single optimization problem or data set.

Supplementary Information The online version contains supplementary material available at https://doi.org/10.1007/978-981-19-5170-1_5.

T. Bartz-Beielstein (✉) · O. Mersmann · S. Chandrasekaran
Institute for Data Science, Engineering and Analytics, TH Köln, Gummersbach, Germany
e-mail: thomas.bartz-beielstein@th-koeln.de

O. Mersmann
e-mail: olaf.mersmann@th-koeln.de

S. Chandrasekaran
e-mail: sowmya.chandrasekaran@th-koeln.de

E. Bartz et al. (eds.), *Hyperparameter Tuning for Machine and Deep Learning with R*, https://doi.org/10.1007/978-981-19-5170-1_5

Single Algorithm Multiple Problem (SAMP): Comparing the results of a single algorithm or learner on many different optimization problems or data sets.

Multiple Algorithm Single Problem (MASP): Comparing the results of multiple algorithms or learners on a single optimization problem or data set.

Multiple Algorithm Multiple Problem (MAMP): Comparing the results of multiple algorithms or learners on many different optimization problems or data sets.

The SASP setting is fundamentally different from the other three settings, because we are not *comparing* results but merely analyzing them. That is, we are evaluating the performance of an *optimization algorithm* $\mathcal{A}$ on a single problem instance π. In the second scenario, we have multiple problem instances $\pi_1, ..., \pi_p$. That means, the second setting is a generalization of the first setting, where we might want to check if our algorithm generalizes to different instances from the same domain or even generalizes to different domains. The third setting generalizes the first by introducing more algorithms $\mathcal{A}_1, ..., \mathcal{A}_a$. Here, we want to compare the performance of these algorithms on a single problem instance and more than likely choose a "best" algorithm. Finally, the last scenario is a combination of the previous two, where we have a algorithms being benchmarked on p problem instances.

For now, we will ignore the challenges posed by the SASP and SAMP settings and focus on the comparison of multiple algorithms. We will denote the random performance measure we use to evaluate an algorithm with Y. Even for deterministic algorithms, it is justified to view this as a random variable since the result still heavily depends on the initial starting parameters, etc. We will assume that we have collected n Independent and Identically Distributed (IID) samples of our performance measure Y for each algorithm and performance metric. These are denoted with $y_1, ..., y_n$.

During all of the following discussions on comparing algorithms, we should always remember that the No Free Lunch theorem (Wolpert and Macready 1997) tells us there is no single best algorithm in both the learning and the optimization setting. We are interested in *comparing* algorithms and *choosing* one that is fit for purpose; we cannot hope to find a single "best" algorithm.

5.2 Ranking

When we are in the MASP setting, there are many established statistical frameworks to analyze the observed performance metrics; see for example Chiarandini and Goegebeur (2010) or Bartz-Beielstein (2015). Here, we will look at a somewhat different approach based on rankings as described in Mersmann et al. (2015). The advantage of ranking-based approaches is their scale invariance.

Consider the case where we have only two algorithms $\mathcal{A}_1$ and $\mathcal{A}_2$. For each algorithm, we observe n values of our performance metric

Algorithm $\mathcal{A}_1$: $y_1^{\mathcal{A}_1}, ..., y_n^{\mathcal{A}_1}$
Algorithm $\mathcal{A}_2$: $y_1^{\mathcal{A}_2}, ..., y_n^{\mathcal{A}_2}$

and we want to decide if $\mathcal{A}_1$ is

1. "better than or equal to" $\mathcal{A}_2$ (denoted by $\mathcal{A}_1 \succ \mathcal{A}_2$);
2. "similar to" $\mathcal{A}_2$ (denoted by $\mathcal{A}_1 \simeq \mathcal{A}_2$);
3. "worse than" $\mathcal{A}_2$ (denoted by $\mathcal{A}_1 \prec \mathcal{A}_2$).

Saying $\mathcal{A}$ is worse than $\mathcal{B}$ is nothing more than saying $\mathcal{B}$ is better than or equal to $\mathcal{A}$:

$$\mathcal{A} \prec \mathcal{B} \iff \mathcal{B} \succ \mathcal{A}.$$

We can also simplify when we consider two algorithms to be similar. We say two algorithms are similar if both are better than or equal to the other one:

$$\mathcal{A} \simeq \mathcal{B} \iff \mathcal{A} \succ \mathcal{B} \land \mathcal{B} \succ \mathcal{A}.$$

Therefore, it is enough to specify the binary relation $\succ$ if we want to decide if some algorithm *dominates* another algorithm. We call $\succ$ the *dominance relation* for our performance metric. One way would be using statistical hypothesis tests as discussed in Sect. 5.6.1 but if n is large,[1] it can be something as simple as the comparison of the mean performance measure attained by each algorithm. It is also possible to think of scenarios where we might be more interested in a consistent result. In these cases, we might compare the variance of the observed performance measures. Finally, if we are really only interested in the absolute best performance the algorithm can deliver, we should compare the minimal or maximal performance measure obtained. For a more detailed description of the different choices available, see Mersmann et al. (2010b). But for now, let's just assume that we are able to define such a dominance relation.

Our dominance relation can have the following useful properties:

reflexive: $\mathcal{A} \succ \mathcal{A}$ for all $\mathcal{A}$ under test. That is, every algorithm is better than or equal to itself. This is a property we want in any dominance relation.
antisymmetric: $\mathcal{A} \succ \mathcal{B} \land \mathcal{B} \succ \mathcal{A} \implies \mathcal{A} \simeq \mathcal{B}$. This is a weaker form of our "similar to" definition above that suffices for our further reasoning.
transitive: $\mathcal{A} \succ \mathcal{B}$ and $\mathcal{B} \succ C$, then $\mathcal{A} \succ C$.
complete: For all distinct pairs of algorithms, either $\mathcal{A} \succ B$ or $\mathcal{B} \succ \mathcal{A}$.

At a minimum, we want our relation to be *reflexive* and *transitive*. We call such a relation a *preorder* and it is the first step toward a relation that induces an order, i.e., gives us a meaningful comparison of all algorithms based on simple pairwise comparisons. Next, we want *antisymmetry* which gives us a *partial order* and finally if the partial order is *complete*, we get a *linear order*. A linear order has quite a few requirements which must be fulfilled. Instead, we could ask ourselves what are the minimum properties we would want? We would certainly want our relation to be *transitive* since otherwise we won't have a ranking, and we also want the relation

---

[1] And see below for reasons why maybe it shouldn't be too large.

to be *complete* so that we can compare all algorithm pairs. An order with just these properties is called a *weak order* and will become important later in our discussion of rankings.

Let's illustrate what we have so far with an example. Assume we have $a = 5$ algorithms and that we measured the performance of each algorithm $n = 15$ times. We can store these results in a 5×15 matrix. Each row stores the results for one algorithm and each column is one observation of the performance measure.

```
t(Y)
```

```
##            A_1         A_2       A_3      A_4       A_5
##  [1,] 11.25802   9.184456 10.91332 9.699683 9.533216
##  [2,] 11.44358   9.227654 11.13609 9.632939 9.339878
##  [3,] 11.49753   9.979770 10.69170 9.411786 9.480409
##  [4,] 11.45181   9.654419 10.87821 9.883699 9.456854
##  [5,] 11.34973   9.359485 10.86492 9.697134 9.416884
##  [6,] 11.67326   9.681364 10.55226 9.586506 9.064084
##  [7,] 11.35203   9.958682 11.04233 9.623864 9.254671
##  [8,] 11.77464  10.094786 11.07630 9.704412 9.397412
##  [9,] 11.35842   9.629653 10.96289 9.586520 9.082003
## [10,] 11.63193   9.664293 11.18674 8.969936 9.164142
## [11,] 11.82829   9.363114 10.88976 9.625530 9.388907
## [12,] 11.09319   9.807767 10.76145 9.495107 9.426344
## [13,] 11.31520   9.587748 11.12167 9.605600 9.365461
## [14,] 11.34429   9.806837 10.87282 9.684919 9.095553
## [15,] 11.49815   9.856715 11.32392 9.848501 9.052752
```

From these raw results, we could derive the incidence matrix of our dominance relation by comparing the mean performance of each algorithm:

```
I <- matrix(0, nrow(Y), nrow(Y))
rownames(I) <- colnames(I) <- rownames(Y)
for (i in 1:nrow(Y)) {
  for (j in 1:nrow(Y)) {
    I[i, j] <- mean(Y[i, ]) >= mean(Y[j, ])
  }
}
I
```

```
##     A_1 A_2 A_3 A_4 A_5
## A_1   1   1   1   1   1
## A_2   0   1   0   1   1
## A_3   0   1   1   1   1
## A_4   0   0   0   1   1
## A_5   0   0   0   0   1
```

And from that the dominance relation using the `relations` package (Meyer and Hornik 2022):

```
r_mean <- relation(incidence = I)
```

We can now check if it is a preorder, partial order, or a linear order:

```
relation_is_preorder(r_mean)
```

```
## [1] TRUE
```

```
relation_is_partial_order(r_mean)
```

```
## [1] TRUE
```

```
relation_is_linear_order(r_mean)
```

```
## [1] TRUE
```

```
relation_is_weak_order(r_mean)
```

```
## [1] TRUE
```

Not surprisingly, we find that the relation is indeed a *linear order*. Using a small helper function, we can pretty print the order

```
show_relation <- function(r) {
  classes <- relation_classes(r)
  class_names <- sapply(
    classes,
    function(x) paste0("{", paste(x, collapse = ", "), "}")
  )
  paste(class_names, collapse = " > ")
}
```

```
show_relation(r_mean)
```

```
## [1] "{A_1} > {A_3} > {A_2} > {A_4} > {A_5}"
```

As expected, algorithm $\mathcal{A}_1$ dominates all other algorithms since it has the highest mean performance of 11.4580052.

Let's see what happens if we use a more nuanced approach using hypothesis tests to derive our dominance relation

```
I <- matrix(0, nrow(Y), nrow(Y))
rownames(I) <- colnames(I) <- rownames(Y)
for (i in 1:nrow(Y)) {
  for (j in 1:nrow(Y)) {
    I[i, j] <- if (i != j) {
      t.test(Y[i, ], Y[j, ],
        paired = TRUE,
        alternative = "less"
      )$p.value > 0.05
    } else {
      1
    }
  }
}
r_ht <- relation(incidence = I)
show_relation(r_ht)
```

```
## [1] "{A_1} > {A_3} > {A_2, A_4} > {A_5}"
```

The resulting dominance relation is not a linear order, because it is not *antisymmetric* since $\mathcal{A}_2 \simeq \mathcal{A}_4$ but $\mathcal{A}_2 \neq \mathcal{A}_4$. It is however still a *weak order* since it is complete and transitive.

```
relation_is_preorder(r_ht)
```

```
## [1] TRUE
```

```
relation_is_partial_order(r_ht)
```

```
## [1] FALSE
```

```
relation_is_linear_order(r_ht)
```

```
## [1] FALSE
```

```
relation_is_weak_order(r_ht)
```

```
## [1] TRUE
```

While a ranking derived from a dominance relation does not give us as many insights as some of the more advanced techniques based on ANOVA or multiple comparison tests, it does extract the essential information we need. From a ranking, we can derive clear preferences for some algorithm or see that a group of algorithms performs similarly.

The real advantage of the ranking-based approach becomes apparent when we leave the MASP setting and go over to the MAMP setting. We can view the MAMP setting as p^2 MASP settings. For each problem instance $\pi_i, \ldots, \pi_p$, we can derive a ranking of the algorithms with the above methodology. This amounts to each problem instance voicing its opinion about which algorithm is preferable. Why is this advantageous when compared to direct performance measure calculations? Because in most cases, the *scale* of our performance measure is specific to the problem instance. We cannot compare the performance measure observed on one problem instance with that on another problem instance. What we can compare are the obtained *ranks*. The ranking is scale-invariant and allows us to aggregate the results of many different MASP scenarios into one MAMP comparison.

5.3   Rank Aggregation

Before we dive into aggregation methods for rankings, let's look at a motivating toy example. An ice cream plant is trying to determine the favorite flavor of ice cream for kids. They let three children rank the flavors based on how well they like them and get the following result:

$$\text{chocolate} \succ \text{vanilla} \succ \text{strawberry} \succ \text{cherry} \succ \text{blueberry}$$
$$\text{vanilla} \succ \text{strawberry} \succ \text{cherry} \succ \text{blueberry} \succ \text{chocolate}$$
$$\text{strawberry} \succ \text{cherry} \succ \text{blueberry} \succ \text{chocolate} \succ \text{vanilla}$$

Here, the children are the "problem instances" and the ice cream flavors are the "algorithms" being ranked. If we simply average the rank for each flavor and then rank the flavors based on this average, we get the following (unsurprising) result:

$$\text{vanilla} \succ \text{strawberry} \succ \text{cherry} \succ \text{chocolate} \succ \text{blueberry} \tag{5.1}$$

Since blueberries are expensive and kids seem to dislike them, they rank last. In fact, we might have suspected that and not taken the flavor blueberry into account. If we remove the blueberry flavor from all three rankings and again calculate the average ranking, we get

$$\text{vanilla} \succ \text{strawberry} \succ \text{chocolate} \succ \text{cherry} \tag{5.2}$$

Notice how deleting the least liked flavor from the list resulted in cherry and chocolate switching positions. Surely, this is not the kind of behavior we would want. But in fact, if we remove the other fruit flavor (cherry), we get an average ranking of

$$\text{vanilla} \simeq \text{strawberry} \simeq \text{chocolate} \tag{5.3}$$

There appears to be no clear preference anymore!

---

[2] Remember p denotes the number of different problem instances in our MAMP setting.

We could also view this as a "fruit conspiracy". Strawberry, cherry, and blueberry, full well knowing that only strawberry has any chance of winning, are in cahoots and all enter the competition. By entering all three fruit flavors the results seem to be skewed in their favor.

We will see that this is an unfortunate side effect of any so-called Consensus Method (CM) for rankings. We have seen in the previous section that, depending on our choice of dominance relation, we arrive at different rankings of our algorithms under test. This is to be expected, since we are ranking them based on different definitions of what we consider to be a better algorithm. We will see that there are different methods for deriving a consensus from our set of rankings and that methods offer different trade-offs between properties that the consensus fulfills. So, before we have seen the first consensus method, we need to accept the fact that from this point forward, we cannot objectively define the best algorithm. Instead, our statement of which algorithm is best depends on our subjective choice of a CM. But not all hope is lost. What we can define are criteria we would want, an ideal CM to have, and then make an informed choice about the trade-off between these criteria.

1. A CM that takes into account all rankings instead of mimicking one predetermined ranking is said to be *non-dictatorial*.
2. A CM that, given a fixed set of rankings, deterministically returns a complete ranking is called a *universal consensus method* or is said to have a *universal domain*.
3. A CM is *independent of irrelevant alternatives*, if given two sets of rankings $R = r_1, \ldots, r_k$ and $S = s_1, \ldots, s_k$ in which for every $i \in 1, \ldots, k$ the order of two algorithms $\mathcal{A}_1$ and $\mathcal{A}_2$ in r_i and s_i is the same; the resulting consensus rankings rank $\mathcal{A}_1$ and $\mathcal{A}_2$ in the same order. Essentially, this means that introducing a further algorithm does not lead to a rank reversal between any of the already ranked algorithms. While this might seem highly desirable (see the above ice cream example), it is also a very strict requirement.
4. A CM which ranks an algorithm higher than another algorithm if it is ranked higher in a majority of the individual rankings fulfills the *majority criterion*.
5. A CM is called *Pareto efficient* if given a set of rankings in which for every ranking an algorithm ai is ranked higher than an algorithm aj, the consensus also ranks ai higher than aj.

No consensus method can meet all of these criteria because the independence of irrelevant alternatives (IIA) and the majority criterion are incompatible. But even if we ignore the majority criterion, there is no consensus method which fulfills the remaining criteria (Arrow 1950). So it is not surprising that if we choose different criteria for our CM, we may get very different consensus rankings.

At this point, we might ask ourselves why bother finding a consensus if it is subjective in the end. And to a certain extent that is true, but it still gives us valuable insights into which algorithms might warrant further investigation and which algorithms perform poorly. However, we have to take care that no accidental or intentional manipulation of the consensus takes place. This can easily happen if the IIA is not fulfilled. Remember how introducing the irrelevant fruit flavors in our toy ice cream example changed the consensus drastically. By adding many similar algorithms or variants of one algorithm, we can skew our analysis and provoke unwanted rank reversals.

Generally, we can differentiate between positional and optimization-based methods. Positional methods calculate sums of scores for each algorithm $\mathcal{A}_i$ over all rankings. The final order is determined by the score obtained by each algorithm. This amounts to

$$\mathcal{A}_i \succ \mathcal{A}_j \iff si \geq sj, \qquad \mathcal{A}_i \simeq \mathcal{A}_j \iff si = sj$$

with the score of algorithm $\mathcal{A}_i$ given by

$$s_i = \sum_{k=1}^{p} s(\mathcal{A}_i, r_{\pi_k}).$$

Here, s denotes a score function and r_{π_k} is the ranking inferred from problem instance π_k. The score function takes as arguments an algorithm and a ranking and returns the score of the algorithm in that ranking.

The simplest score function we might use assigns a value of one to the best algorithm in each ranking while all other algorithms get a value of zero. Although this is somewhat intuitive, undesirable consensus rankings can occur. Consider the situation with two different rankings of three algorithms:

$$\mathcal{A}_1 \succ \mathcal{A}_2 \succ \mathcal{A}_3 \quad \text{and} \quad \mathcal{A}_3 \succ \mathcal{A}_2 \succ \mathcal{A}_1.$$

Using the above score function, we would obtain the following scores:

$$s_1 = 1 + 0 = 1 \quad s_2 = 0 + 0 = 0 \quad s_3 = 0 + 1 = 1$$

which leads to the consensus ranking

$$\{\mathcal{A}_1 \simeq \mathcal{A}_2\} \succ \mathcal{A}_2.$$

This is counterintuitive since the two rankings are opposed and we'd expect them to cancel out and give

$$\{\mathcal{A}_1 \simeq \mathcal{A}_2 \simeq \mathcal{A}_2\}.$$

The Borda count method (de Borda 1781) solves this issue and assigns an algorithm one point for each algorithm that is not better than

$$s^{BC}(\mathcal{A}_i, r) = \sum_{i \neq j} \mathbf{I}(\mathcal{A}_i \succ \mathcal{A}_j).$$

In the case of no ties, it reduces the ranks of the data. For our example rankings above, we get

$$s_1 = 2 + 0 = 2 \quad s_2 = 1 + 1 = 2 \quad s_3 = 0 + 2 = 2$$

and the consensus ranking

$$\{\mathcal{A}_1 \simeq \mathcal{A}_2 \simeq \mathcal{A}_2\}$$

which is more intuitive than the previous result. Unfortunately, the Borda method does not fulfill the majority or the IIA criterion. It is still a popular consensus method because it can be easily implemented and understood. The main criticism voiced in the literature is that it implicitly, like all positional consensus methods, assumes a distance between the positions of a ranking.

A completely different approach is to frame the CM as an optimization problem where we want to find a ranking that minimizes a function of the distances to all of the individual rankings. Cook and Kress (1992) give a gentle introduction to this line of thought and present a wide variety of possible distance functions. Central to this is a notion of betweenness, expressed by pairwise comparisons. Here, we will focus on the axiomatically motivated symmetric difference distance function[3] originally proposed by Kemeny and Snell (1962), but the general procedure is the same regardless of the distance function chosen. First, we pick a set C of admissable consensus rankings. This could be the set of all *linear* or *weak orderings* of our algorithms. Then, we solve the following optimization problem:

$$\arg\min_{c \in C} L(c) = \arg\min_{c \in C} \sum_{i=1}^{p} d(c, r_{\pi_i})^{\ell}, \quad \ell \geq 1.$$

Setting $\ell = 1$ results in what is called a median consensus ranking and $\ell = 2$ results in a mean consensus ranking.

Let's revisit the ice cream example and see what the consensus is according to Borda or using the symmetric difference.

```
show_relation(child1)

    ## [1] "{chocolate} > {vanilla} > {strawberry} > {cherry} > {blueberry}"

show_relation(child2)

    ## [1] "{vanilla} > {strawberry} > {cherry} > {blueberry} > {chocolate}"

show_relation(child3)

    ## [1] "{strawberry} > {cherry} > {blueberry} > {chocolate} > {vanilla}"
```

---

[3] The symmetric difference counts the number of cases where $\mathcal{A}_i \succ \mathcal{A}_j$ is contained in one of the relations but not the other.

The Borda consensus among the three children is

```
ranks <- relation_ensemble(child1, child2, child3)
r_borda <- relation_consensus(ranks, "Borda")
show_relation(r_borda)

  ## [1] "{strawberry} > {vanilla} > {cherry} > {chocolate} > {blueberry}"
```

and the symmetric difference-based consensus among all linear orderings of the flavors is

```
r_sd <- relation_consensus(ranks, "symdiff/L")
show_relation(r_sd)

  ## [1] "{vanilla} > {strawberry} > {cherry} > {blueberry} > {chocolate}"
```

We see that the Borda consensus falls into the "fruit-gang trap" and ranks the strawberry flavor first. The symmetric difference-based consensus on the other hand ranks vanilla higher than strawberry because in two out of three rankings, it ranks higher than strawberry.

Unfortunately, we cannot give a general recommendation regarding the introduced consensus methods as each method offers a different trade-off of the consensus criteria (Saari and Merlin 2000). The symmetric difference combined with linear or weak orderings meet the majority criterion and thus cannot meet the IIA criterion simultaneously. However, on real data, as seen in the ice cream example, they rarely result in rank reversals if algorithms are added or dropped. The Borda count method does not fulfill either of these criteria. Saari and Merlin (2000) however showed that both methods always rank the respective winner above the loser of the other method.

Finally, it is important to note that consensus rankings generally do not admit nesting in a hierarchical structure. For example, separate consensus rankings could be of interest for problem instances with specific features. While this certainly is a valid and meaningful approach, one has to keep in mind that an overall consensus of these separate consensus rankings does not necessarily have to equal the consensus ranking directly generated based on all individual rankings.

5.4 Result Analysis

Many of the Machine Learning (ML) and Deep Learning (DL) methods are stochastic in nature as there is randomness involved as a part of optimization or learning. Hence, these methods could yield different results to the same data for every run. To access the performance of the model, one single evaluation may not be sufficient. To statistically evaluate the variance of the obtained results, multiple repeats have to be performed and the summary statistics of the performance measure are to be reported.

Generally, the performance of the ML and DL methods can be analyzed considering model quality and runtime. The model quality is determined using the Root Mean

Squared Error (RMSE) for the regression models and the Mean Mis-Classification Error (MMCE) for the classification models as discussed in Sect. 2.2.

Often, these quality metrics are compared among different algorithms to analyze their performances. Hence, the tuners aim to minimize these metrics. As these metrics can be affected by the algorithm's and tuner's stochastic nature, the experiment has to be repeated for a specific number of times. It enables better estimation of the model quality parameter using descriptive and Exploratory Data Analysis (EDA) tools. Also, statistical inference is highly recommended in understanding the underlying distribution of the model quality parameters.

EDA is a statistical methodology for analyzing data sets to summarize their main characteristics (Tukey 1977; Chambers et al. 1983). The EDA tools are employed to analyze and report the performance of the ML models. This includes both descriptive and graphical tools. The numerical measures include reporting the *mean*, *median*, *best*, *worst*, and *standard deviation* of the performance measures of the algorithms obtained for certain number of repeats. They measure the central tendency and the variability of the results. The graphical tools like histograms, and box and violin plots provide information about the shape and the distribution of the performance measures, respectively. These statistics are necessary, but are not always sufficient to evaluate the performances. Kleijnen (1997), Bartz-Beielstein et al. (2010), Myers et al. (2016), Montgomery (2017), and Gramacy (2020) are good starting points. More information about various techniques and best practices in analyzing the performance measures can be found in Bartz-Beielstein et al. (2020b).

5.5   Statistical Inference

Statistical inference means drawing conclusions from partial information of a population about the whole population using methods based on data analysis and probability theory. Statistical inference is recommended in making decisions about identifying the best algorithm and tuner. The key ingredient of statistical inference is hypothesis testing (Neyman 1950). As a part of pre-data analysis, the null hypothesis H_0 can be formulated as "There is no statistically significant difference between the compared algorithms", while the alternative hypothesis H_1 states that there exists a statistically significant difference between the compared algorithms. Hypothesis testing will be outlined in Sect. 5.6.1.

The hypothesis testing can be classified into *parametric* and *non-parametric* tests. For the case of parametric tests, the distributional assumptions have to be satisfied, one of which is Normal, Independent and Identically Distributed (NIID) data. If the distributional assumptions are not met, non-parametric tests are employed. For the case of single pairwise comparison, the most commonly used parametric test is the t-test (Sheskin 2003) and its non-parametric counter-part is the Wilcoxon-rank sum test (Hart 2001). And in case of multiple comparisons, one commonly used parametric test is the one-way *ANOVA* (Lindman 1974), while its non-parametric test counter-part is the *Kruskal-Wallis rank sum test* (Kruskal and Wallis 1952). The following sections analyze parametric tests.

5.6 Definitions

5.6.1 Hypothesis Testing

Generally, hypothesis testing can be either one-sided or two-sided: $H_0 : \tau \leq 0$ versus $H_1 : \tau > 0$ (one-sided) or $H_0 : \tau = 0$ versus $H_1 : \tau \neq 0$ (two-sided), where H_0 and H_1 denote the corresponding hypotheses that will be explained in this section. For the purpose of performance comparison of the two methods, we consider a one-sided test and question whether method $\mathcal{A}$ is better than method $\mathcal{B}$. Let $p(\mathcal{A})$ and $p(\mathcal{B})$ represent the performance of method $\mathcal{A}$ and $\mathcal{B}$, respectively. If we consider a minimization problem, the smaller the values the better the performance of the method. For method $\mathcal{A}$ to be better than method $\mathcal{B}$, $p(\mathcal{A}) < p(\mathcal{B}) \Leftrightarrow p(\mathcal{B}) - p(\mathcal{A}) > 0 \Leftrightarrow \tau > 0$.

To state properties of the hypothesis, the symbol μ will be used for the mean, whereas the symbol τ denotes the difference between two means. For example, $\tau = \mu_1 - \mu_0$ or variations of the mean, e.g., $\tau = \mu + \Delta$.

Definition 5.2 (*One-sided Hypothesis Test*) The hypothesis is then formulated as

$$H_0 : \tau \leq 0 \text{ versus } H_1 : \tau > 0, \tag{5.4}$$

where τ denotes the range of possible values.

Definition 5.3 (*Test Statistic*) The test statistic $d(Y)$ reflects the distance from H_0 in the direction of H_1. Assuming the data follow a normal distribution, i.e., $Y \sim \mathcal{N}(\mu_0, \sigma^2)$, the test statistic reads

$$d(Y) = \sqrt{n}(\bar{Y} - \mu_0)/\sigma. \tag{5.5}$$

In the remainder of this chapter, we assume that data are NIID.

Definition 5.4 (*Cut-off Point: $c_{1-\alpha}$*) The $c_{1-\alpha}$ is a threshold value or the cut-off point.

Definition 5.5 (*Upper-tail of the Standard Normal Distribution: $u_{1-\alpha}$*) $u_{1-\alpha}$ denotes the value of the normal distribution which cuts off the upper-tail probability of α.

Based on the test statistic from Eq. 5.5, we can calculate the cut-off point $c_{1-\alpha}$: $d(Y) = \sqrt{n}(\bar{Y} - \mu_0)/\sigma = u_{1-\alpha} \Leftrightarrow \bar{Y} = \mu_0 + (u_{1-\alpha})\sigma/\sqrt{n} = c_{1-\alpha}$. When a test statistic is observed beyond the cut-off point, $d(Y) > c_{1-\alpha}$, we reject the H_0 at a significance level α. Otherwise the H_0 is not rejected.

This hypothesis test can lead to two kinds of errors based on the decision taken.

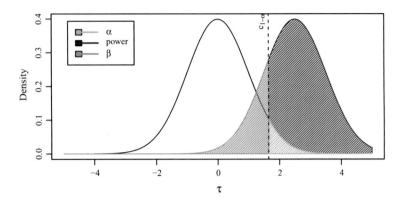

Fig. 5.1 Hypothesis test

Definition 5.6 (*Type I and II Errors*) They are the Type I and the Type II errors, which are pre-specified before the experiment is carried out.

1. A Type I error occurs while incorrectly rejecting the null hypothesis when it is true. The probability of committing a Type I error is called the significance level and is denoted as α. In other words, α is the acceptable probability for Type I error to occur, which is decided by the user. The Type I error can be represented as shown in Fig. 5.1. $\alpha = P_{H_0}(d(Y) > c_{1-\alpha})$.
2. A Type II error occurs while incorrectly rejecting the alternative hypothesis, when it is true: $\beta = P_{H_1}(d(Y) \leq c_{1-\alpha})$.

The notation $P_H(y)$ represents the probabilistic assignments under a model, i.e., the probability of y under the hypothesis H. The power $(1 - \beta)$ is the probability of correctly rejecting the null hypothesis when it is false.

Definition 5.7 (*Paired Samples*) Two samples X_1 and X_2 are considered *paired*, if there is a relation that assigns each element in X_1 uniquely to one element in X_2.

Example: Paired Samples

Therefore, we consider results from running deterministic optimization methods $\mathcal{A}$ and $\mathcal{B}$ paired, if they are using the same starting points (the starting points can be used for indexing the sample points). The starting points are randomly generated, using the same seed for each sample.

Example: Conjugate Gradient versus Nelder-Mead

We will consider the performance differences between two optimization methods. To enable replicability, we have chosen two optimization methods (optimizers) that are available "out of the box" in every R installation via the optim function. They are described in the R help system as follows (R Core Team 2022):

1. Method Conjugate Gradient (CG) is a conjugated gradients method based on Fletcher and Reeves (1964). Conjugate gradient methods will generally be more fragile than the Broyden, Fletcher, Goldfarb, and Shanno (BFGS) method, but as they do not store a matrix they may be successful in much larger optimization problems.
2. Method Nelder and Mead Simplex Algorithm (NM) uses only function values and is robust but relatively slow (Nelder and Mead 1965). It will work reasonably well for non-differentiable functions.

CG and NM will be tested on the two-dimensional Rosenbrock function (Rosenbrock 1960). The function is defined by

$$f(x_1, x_2) = (1 - x_1)^2 + 100(x_2 - x_1^2)^2. \tag{5.6}$$

It has a global minimum at $(x_1, x_2) = (1, 1)$. To keep the discussion focused, assume that results from $n = 100$ runs of each method are available, i.e., in total, 200 runs were performed. Let $y_{i,j}$ denote the result of the jth repetition of the ith method, i.e., the vector $y_{1,.}$ represents 100 results of the CG runs.

We will consider the performance differences $d_j = y_{1,j} - y_{2,j}, \quad j = 1, \ldots, n$, with corresponding mean $\bar{d} = 9.02$. Based on

$$S_d = \left(\frac{\sum_{j=1}^n (d_j - \bar{d})^2}{n - 1} \right)^{1/2} \tag{5.7}$$

we can calculate the sample standard deviation of the differences as $S_d = 30.73$.

As $\bar{d}$ is positive, we can assume that method NM is superior. We are interested to see whether the difference in means is smaller or larger than μ_0 and formulate the test problem as

$$H_0 : \mu \leq \mu_0 \text{ versus } H_1 : \mu > \mu_0,$$

in our case: $\mu_0 = 0$. And, if H_0 is rejected then it signifies that NM outperforms CG for the given test function.

We will use the test statistic as defined in (5.5) which follows a standard normal distribution if H_0 is true ($\mu \leq \mu_0$). Then

$$P \left(\frac{\overline{Y} - \mu_0}{\sigma/\sqrt{n}} > u_{1-\alpha} \right) \leq \alpha, \text{ otherwise } P \left(\frac{\overline{Y} - \mu_0}{\sigma/\sqrt{n}} > u_{1-\alpha} \right) > \alpha, \tag{5.8}$$

where $u_{1-\alpha}$ denotes the cut-off point; see Definition 5.5.

The test $T(\alpha)$ results in rejecting the null hypothesis H_0 if $d(y) > u_{1-\alpha}$ and in not rejecting H_0 otherwise. For $\alpha = 0.025$ and $u_{1-\alpha} = 1.96$, we get $d(y) = (\overline{y} - \mu_0)/(\sigma/\sqrt{n}) = 2.93 > 1.96 = u_{1-\alpha}$, i.e., H_0 will be rejected.

A sample size of $n = 100$ was chosen without any statistical justification: it remains unclear whether ten samples might be sufficient or whether one thousand samples should have been used. The power calculation, which will be discussed next, provides a proven statistical tool to determine adequate sample sizes for planned experimentation.

5.6.2 Power

The power function that is used to calculate the power for several alternatives μ_1 is defined as

Definition 5.8 (*Power Function*)

$$\text{Pow}(\mu_1) = P_{\mu=\mu_1}\left(\frac{\overline{Y} - \mu_0}{\sigma/\sqrt{n}} > u_{1-\alpha}\right) = 1 - \Phi\left(u_{1-\alpha} - \frac{\mu_1 - \mu_0}{\sigma/\sqrt{n}}\right) \tag{5.9}$$

where $\mu_1 = \mu_0 + \Delta$ and Δ denotes the relevant difference.

In our example, we set up a one-sided test with $H_0 : \mu_0 = 0$ and the following parameters:

1. significance level: $\alpha = 0.025$
2. beta (1-power): $\beta = 1 - 0.8 = 0.2$
3. relevant difference: $\Delta = 10$
4. between-sample standard deviation: $\sigma = 30.73$.

The relationship between power and sample size is illustrated in Fig. 5.2.

5.6.3 p-Value

The p-value quantifies how strongly the data contradicts the null hypothesis, and it allows others to make judgments based on the significance level of their choice (Mayo 2018; Senn 2021).

Definition 5.9 (*p -value*) A p-value is the probability of observing an outcome as extreme or more extreme than the observed outcome $\overline{y}$ if the null hypothesis is true. It is defined as the α' value with

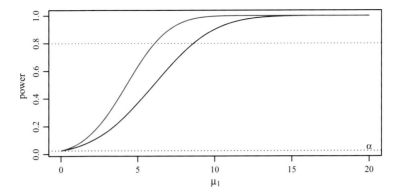

Fig. 5.2 Power for n = 100 (black) and n = 200 (red) for varying μ_1 values. This figure illustrates that larger sample sizes result in higher power

$$d(Y) > u_{1-\alpha'} \Leftrightarrow \alpha' = 1 - \Phi\left(\sqrt{n}(\bar{y} - \mu_0)/\sigma\right),$$

under the assumption that H_0 is true.

If an effect τ measures the true difference between the performance of two methods and y is a statistic used to measure the difference between methods, a one-sided p-value can be defined as

$$P_{\tau=0}(y \geq \bar{y}) \tag{5.10}$$

where $\bar{y}$ is the observed value of the statistic if H_0 is true. The p value can be used for going beyond the simple decision *reject* or *not reject*.

Senn (2002) claims that p-values are a perfectly reasonable way for scientists to communicate the results of a significance test, even when making decisions rather than conclusions. Small p-values indicate that either H_0 is not true or a very unlikely event has occurred (Fisher 1925).

Example: CG versus NM continued

Considering the CG versus NM example (Sect. 5.6.1), the observed difference $\bar{d} = 9.02$, and the corresponding p-value of 0.0017 is obtained.

5.6.4 Effect Size

The effect size is an easy scale-free approach to quantifying the size of the performance difference between the two methods.

Definition 5.10 (*Effect size*) The *effect size* is the standardized mean difference between the two methods, say $\mathcal{A}$ and $\mathcal{B}$ (Cohen 1977):

$$\text{Cohen's } d = \frac{\bar{y}_{\mathcal{A}} - \bar{y}_{\mathcal{B}}}{S_p} \tag{5.11}$$

$$S_p = \sqrt{\frac{(n_{\mathcal{B}} - 1)s_{\mathcal{B}}^2 + (n_{\mathcal{A}} - 1)s_{\mathcal{A}}^2}{n_{\mathcal{A}} + n_{\mathcal{B}} - 2}}, \tag{5.12}$$

where $\bar{y}_{\mathcal{A}}$ and $\bar{y}_{\mathcal{B}}$ is the sample mean of the method $\mathcal{A}$ and $\mathcal{B}$, respectively. The S_p is the pooled standard deviation, $n_{\mathcal{A}}, n_{\mathcal{B}}$ are the sample size of each method, and $s_{\mathcal{A}}, s_{\mathcal{B}}$ are the standard deviation of each method. As a guideline, Cohen suggested effect size as small (0.2), medium (0.5), and large (0.8) but with a strong caution to the applicability in different fields.

Hedges and Olkin (1985) identified that Cohen's d is biased and it slightly overestimates the standard deviation and introduced a correction measure as

$$\text{Hedge's } g = 1 - \frac{3}{4(n_{\mathcal{A}} + n_{\mathcal{B}}) - 9} \times \text{Cohen's } d. \tag{5.13}$$

Example: CG versus NM continued

Again, considering the CG versus NM example (Sect. 5.6.1), Cohen's d and Hedge's g, which are the standardized mean difference between the two methods, can be calculated using (5.11) and (5.13) as $d = 0.415$ and $g = 0.4134$, respectively. Both values indicate that the observed mean difference is of a smaller magnitude.

5.6.5 Sample Size Determination and Power Calculations

Adequate sample size is essential for comparing algorithms. Even for deterministic optimizers, it is recommended to perform several runs with varying starting points instead of using results from one run of each algorithm. But "the more the merrier" is not efficient in this context, because additional runs incur additional costs. Statistical inference provides tools for tackling this trade-off between cost and effectiveness.

5.6.5.1 Five Basic Factors

The usual point of view is that the sample size is the determined function of variability, statistical method, power, and difference sought. We consider a one-sided test as defined in Eq. 5.4.

Definition 5.11 (*Five basic factors*) While discussing sample size requirements, Senn (2021) introduced the following conventions regarding symbols:

α: the probability of a type I error, given that the null hypothesis is true.

β: the probability of a type II error, given that the alternative hypothesis is true.
Δ: the difference sought. In most cases, one speaks of the "relevant difference" and this in turn is defined "as the difference one would not like to miss". Notation: In hypothesis testing, Δ denotes a particular value within the range of possible values τ.
σ: the presumed standard deviation of the outcome.
n: the number of runs of each method. Because two methods are compared; the total number is $2 \times n$.

5.6.5.2 Sample Size

Based on the definition of the type II error rate for $1 - \beta$ for μ_1, the sample size can be calculated for the type II error rate, i.e.,

$$\Phi\left(u_{1-\alpha} - \frac{\mu_1 - \mu_0}{\sigma/\sqrt{n}}\right) = \beta \Leftrightarrow n = \frac{\sigma^2}{(\mu_1 - \mu_0)^2}(u_{1-\alpha} - u_\beta)^2 = \frac{\sigma^2}{\Delta^2}(u_{1-\alpha} - u_\beta)^2,$$

which gives an estimate of the required sample size $n = n(\alpha, \beta, \sigma, \mu_0, \mu_1) = n(\alpha, \beta, \sigma, \Delta)$.

Any four factors from Definition 5.11 are enough to determine the fifth factor uniquely. First, we consider the formula for sample size, n as a function of α, β, Δ, and σ. For a one-sided test of size α, the (approximate) formula for sample size is

$$n \approx 2 \times (u_{1-\alpha} + u_{1-\beta})^2 \sigma^2 / \Delta^2, \tag{5.14}$$

where $u_{1-\alpha}$ denotes the value of the normal distribution which cuts off the upper-tail probability of α.

Hence, for the CG versus NM example (Sect. 5.6.1), if the relevant difference is $\Delta = 10$ then approximately 148 completing runs per method are required.

Example: Sample size determination

Compare two optimization methods, say $\mathcal{A} = CG$ and $\mathcal{B} = NM$. Therefore, we set up a one-sided test with the following parameters:

1. significance level: $\alpha = 0.05$
2. beta (1-power): $\beta = 1 - 0.8 = 0.2$
3. relevant difference: $\Delta = 200$
4. between-sample standard deviation: $\sigma = 450$.

We will use the function `getSampleSize` from the R package `SPOT` to determine the sample size n. All calculations shown in this chapter are implemented in this package.

```
library("SPOT")
nsamples <- round(getSampleSize(
   mu0 = 0, mu1 = 200,
   alpha = 0.05, beta = 0.2,
   sigma = 450,
   alternative = "one.sided"
), 0)
```

Based on Eq. 5.14, approximately $n = 63$ completing runs per method are required.

Although sample size calculation appears to be transparent and simple, there are several issues with this approach that will be discussed in the following.

5.6.6 Issues

In this section, we will consider issues with sample size determination, with power calculations, and with hypotheses and wrong conclusions from hypothesis testing. Our presentation (and especially the examples) is based on the discussion in Senn (2021).

Issues with sample size determination can be caused by the computation of the standard deviation, σ: This computation is a chicken or egg dilemma, because the between-sample standard deviation will be unknown until the result of the experiment is known. But the experiment must be planned before it can be run. Furthermore, Eq. 5.14 is only an approximate formula, because it is based on the assumption that the standard deviation is known. The experiments we use are based on using an estimate obtained from the examined sample.

There is no universal standard for a relevant difference Δ. This creates another problem in determining sample size, since significant differences are application-dependent.

Errors can cause issues with sample size determination, because the levels of α and β are relative: α is an actual value used to determine significance in analysis, while β is a theoretical value used for planning (Senn 2021). Frequently, the error values are chosen as $\alpha = 0.05$ and $\beta = 0.20$. However, in some cases, the value of β ought to be much lower, but if only a very small number of experiments are feasible, a very low value of β might not be realistic. The same considerations are true for α, because α and β cannot be reduced simultaneously without increasing the sample size.

In practice, sample size calculation might be flawed. For example, $n = 10$ or $n = 100$ are popular sample sizes, but they are often chosen without any justification. Some authors justify their selection by claiming that "this is done by everyone".

In some situations, there is enough knowledge to plan an experiment, i.e., the number of experiments to be performed is known. Nuclear weapons tests are an extreme example of this situation.

Furthermore, Senn (2021) claims that the sample size calculation can be "an excuse for a sample size and not a reason". In practice, there is a usually undesirable tendency to "adjust" certain factors, notably the difference sought and sometimes the power, in light of *practical sample size requirements.*

Tip: Sample Size Determination

Perform pre-experimental runs to compute the (approximate) sample size before the full experiment is started.

In addition to issues with sample size determination, also issues with power calculations might arise. The fact that a sample size has been chosen which seemingly has 80% power does not guarantee that there is an 80% chance that there is an effect (alternative H_1 is true) (Senn 2021). Even if the whole experimental setup and process are correct, external failures can happen and that is outside of the experimenter's control: The methods or the algorithm may not work. Importantly, if an algorithm does not work we must recognize this; see the example in Sect. 5.8.2.1. But even if the algorithm is successful, it may not produce a relevant difference. Or, looking at another extreme, the algorithm might be better than planned for—so the sample size could have been chosen smaller. In addition, experimental errors might occur that are not covered by the assumptions made for the power (sample size) calculation. The calculations are made under the assumption that the experiment is performed correctly. Or, as Senn (2021) states: Sample size calculation does not allow for "acts of God" or dishonest or incompetent investigators. Thus, although we can affect the probability of success by adjusting the sample size, we cannot fix it.

Finally, there are issues with hypotheses and wrong conclusions based on hypothesis testing. Selecting the correct hypothesis pair, e.g., $H_0 : \tau \leq 0$ versus $H_1 : \tau > 0$ (one-sided) or $H_0 : \tau = 0$ versus $H_1 : \tau \neq 0$ (two-sided) is not always obvious.

In the context of clinical testing, Senn (2021) states that the following statement is a *surprisingly widespread piece of nonsense*:

> If we have performed a power calculation, then upon rejecting the null hypothesis, not only may we conclude that the treatment is effective but also that it has a clinically relevant effect.

Consider, for example, the comparison of an optimization method, $\mathcal{A}$ with a second one, say method $\mathcal{B}$, based on a two-sided test. Let τ be the true difference in performance ($\mathcal{A}$ versus $\mathcal{B}$). We then write the two hypotheses,

$$H_0 : \tau = 0 \text{ versus } H_1 : \tau \neq 0. \tag{5.15}$$

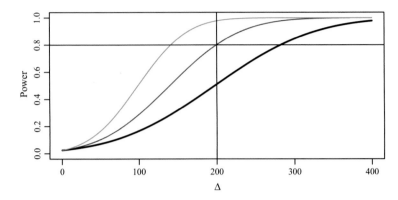

Fig. 5.3 Power as a function of the relevant difference Δ for a two-parallel-group experiment (black = 40, red = 80, and green = 160 runs). If the relevant difference Δ is 200, n = 80 runs per method are needed for 80% power

By rejecting the null hypothesis, we are in favor of the alternative, H_1, which states that there is a non-zero difference. The sign of this difference might indicate whether $\mathcal{A}$ is superior or inferior to $\mathcal{B}$.

Replacing Eq. 5.15 with

$$H_0 : \tau = 0 \text{ versus } H_1 : \tau \geq \Delta \qquad (5.16)$$

would imply that we know one algorithm is better than the other *before* the experiments are performed. But this is usually not known prior to the experiment—the whole point of the experiment is to determine which algorithm performs better. Therefore, we will consider a one-sided test as specified in Eq. 5.4. This procedure will be exemplified in Sect. 5.8.

We have highlighted some important issues with sample size determination, power calculations, and hypotheses tests. Senn (2021) mentions many more, and the reader is referred to his discussion.

Tips

Plotting the power function for an experiment is recommended. This is illustrated in Fig. 5.3.

Last but not the least, issues with the "large n problem", i.e., the topic "large versus small samples", should be considered. Senn (2021), Sect. 13.2.8 states:

1. other things being equal, significant results are more indicative of efficacy if obtained from large experiments rather than small experiments.

2. But consider: if the sample size is increased, not only is the power of finding a relevant difference, Δ, increased, but the smallest detectable difference also decreases.

5.7 Severity

5.7.1 Motivation

Severity has been proposed as an approach to tackle the issues discussed in Sect. 5.6.6 by philosopher of science Mayo (1996). To explain the concept of severity, we start with an example that was taken from (Senn 2021).

Example: High Power

Consider an algorithm comparison using a one-sided test with $\alpha = 0.025$ but with a very high power, say 99%, for a target relevant difference of $\Delta = 200$. The standard deviation of the differences in the mean is taken to be 450. Note, except for drastically reducing the error of the second kind from $\beta = 0.2$ down to $\beta = 0.01$, this example is similar to the Example "Sample Size Determination" in Sect. 5.6.5. A one-sided hypothesis test as specified in Eq. 5.4 with the following parameters is performed:

1. significance level: $\alpha = 0.025$
2. power: $1 - \beta = 0.99$
3. relevant difference: $\Delta = 200$
4. between-sample standard deviation: $\sigma = 450$.

A standard power calculation, see Eq. 5.14, suggests $n \approx 186$ samples for each configuration, which we round up to $2 \times 200 = 400$ in total. This value gives a standard error for the difference of $450 \times \sqrt{2/200} = 45$.

We run the experiments (assuming unpaired, i.e., independent samples) and the result is significant, i.e., we have observed a difference of $\bar{y} = 90$. We get the p-value 0.0231.

How can we interpret the results from this experiment, e.g., the p-value? Although the p-value of 0.0231 is statistically significant, i.e., p-value $< \alpha$, we cannot conclude that the H_1 is true. The probability of occurrence of a type I error has to be acknowledged. The situation is shown in Fig. 5.4. Observing a $\bar{y} = 90$ is more likely under H_0 than under H_1. This is evident by comparing the height of the density curve at $\bar{y} = 90$ both under the H_0 and H_1, respectively. Hence, this is more likely to be the case of a type I error. Although the power is relatively high $(1 - \beta = 0.99)$, it

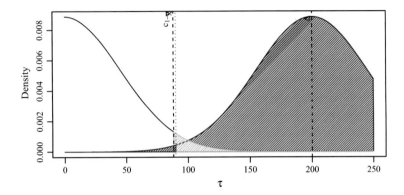

Fig. 5.4 Severity (red), type I error rate (gray), and power (blue). Since $\bar{y}$ is larger than $c_{1-\alpha}$, the null hypothesis is rejected

would be an error to claim that the experiment has an effect $\Delta = \mu_1 - \mu_0 = 200$.[4] There are two reasons:

1. This test did not make extensive use of $\bar{y}$, the actual difference observed. The actual difference observed is only used to calculate the test statistic and to decide whether the null hypothesis should be rejected.
2. Going beyond the simple decision *reject* or *not reject*, the p value can be used. The actual difference observed, $\bar{y} = 90$, is closer to 0 than to 200. Because 90 is farther away from 200 than from 0, this is far from good evidence that the true difference is as large as the relevant difference of 200.

Senn (2021) proposed ways for solving this problem, e.g., using a so called *point-estimate* of the true difference together with associated confidence limits, or using an irrelevant difference approach, or using severity.

5.7.2 Severity: Definition

Severity is a measure of the plausibility of a result which considers the decision and the data: after the decision is made, the severity of rejecting or not rejecting the null hypothesis can be calculated. It uses post-data information and provides means for answering the question:

How can we ensure that the results are not only statistically but also scientifically relevant?

The concept of *Severity* was introduced by Mayo and Spanos (2006) (see also Mayo 2018):

---

[4] Note: $\mu_0 + \Delta = \mu_1$.

Table 5.1 Power $(1 - \beta)$, significance level (α), p-value, and severity. P_{H_0} denotes the probability under the assumption that H_0 is true, whereas P_{H_1} denotes the probability under the assumption that H_1 is true

$1 - \beta$	α	p-value	Severity
$P_{H_1}(Y > c_{1-\alpha})$	$P_{H_0}(Y > c_{1-\alpha})$	$P_{H_0}(Y > \bar{y})$	$S_{nr}: P_{H_1}(Y > \bar{y})$
			$S_r: P_{H_1}(Y \leq \bar{y})$
$P_{H_1}(d(Y) > u_{1-\alpha})$	$P_{H_0}(d(Y) > u_{1-\alpha})$	$P_{H_0}(d(Y) > d(\bar{y}))$	$S_{nr}:$
			$P_{H_1}(d(Y) > d(\bar{y}))$
			$S_r: P_{H_1}(d(Y) \leq d(\bar{y}))$

The result that hypothesis H is not rejected is severe only if it is very unlikely that this result will also occur if H is false.

Severity offers a meta-statistical principle for evaluating the proposed statistical conclusions. It shows how well-tested (not how likely) hypotheses are. It is therefore an attribute of the entire test procedure. The severity of the test and the resulting outcome can be evaluated.

Definition 5.12 (*Severity*) Severity is defined separately for the non-rejection (S_{nr}) and the rejection (S_r) of the null hypothesis as in (5.17).

$$S_{nr} = P_{H_1}(d(Y) > d(y))$$
$$S_r = P_{H_1}(d(Y) \leq d(y)). \tag{5.17}$$

The S_{nr} values increase monotonically from 0 to 1 as a function of τ. The S_r values decrease monotonically from 1 to 0 as a function of τ. The closer the value is to 1, the more reliable is the decision made with the hypothesis test. The key difference between power and severity is that severity depends on the data and the test statistic, i.e., $d(y)$ instead of $c_{1-\alpha}$.

The severity is an analogous probability to Eq. 5.10 that considers non-zero τ values. The severity of rejection, which considers values in the other direction, $y \leq \bar{y}$ is calculated as

$$S_r(\tau') = P_{\tau=\tau'}(y \leq \bar{y}), \tag{5.18}$$

if H_0 is rejected and $S_{nr}(\tau') = P_{\tau=\tau'}(y \geq \bar{y})$, otherwise. Table 5.1 shows the relations between power $(1 - \beta)$, significance level (α), p-value, and severity.

Example: High Power (Continued)

Figure 5.5 plots the severity for the given example against every possible value of the true difference in the performance τ (Senn 2021).

Labeled on the graph are values of $\tau = \bar{y}$, the observed difference, for which the severity is 0.5, and $\tau = \Delta$, the value used for planning. The severity of rejecting H_0 is only 0.0075 for this value. Figure 5.5 exhibits that $\tau > 200$ has a very low severity.

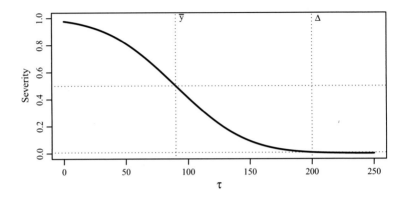

Fig. 5.5 Severity of rejecting H_0, S_R as a function of $\Delta = \mu_1 - \mu_0$. $S_R(0) = 1 - p$

The p-value, here 0.0231, is smaller than $\alpha = 0.025$. Note, in case of rejection H_0, severity is $1 - p$ for $\Delta = \mu_1 - \mu_0 = 0$. The severity of not rejecting the null hypothesis is the same as the p-value for $\Delta = 0$.

Example: Conjugate Gradient versus Nelder-Mead (Continued)

Let us now revisit the CG versus NM example (Sect. 5.6.1) and calculate the severity. Given the observed difference 9.02 and sample size n=100, the decision based on the p-value of 0.0017 is to reject H_0. Considering a target relevant difference of $\Delta = 10$, the severity of rejecting H_0 is 0.37 and is shown in the left panel in Fig. 5.6. The right panel in Fig. 5.6 shows the severity of rejecting H_0 as a function of τ. Based on the result of the hypothesis test for the given data, NM seems to outperform CG. And, claiming that the true difference is as large as or larger than 10 has a very low severity, whereas differences smaller than 7 are well supported by severity.

5.7.3 Two Examples

We will use two illustrative examples for severity calculations that are based on the discussions in Mayo (2018), Bönisch and Inderst (2020), and Senn (2021). In each example, 100 samples from a $\mathcal{N}(\mu, \sigma^2)$-distributed random variables are drawn, but with different means. The first example represents a situation in which the true difference is small compared to the variance in the data, whereas the second example represents a situation in which the difference is relatively large. The first example uses the sample mean $\mu_1 = 1e - 6$ (data set I), the second sample $\mu_2 = 3$ (data set

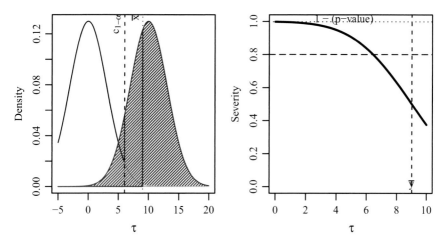

Fig. 5.6 Left: Severity of rejecting H_0 (red), power (blue) for a target relevant difference $\Delta = 10$. Right: Severity of rejecting H_0 as a function of τ

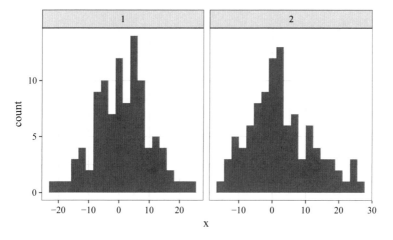

Fig. 5.7 Data sets I and II. Histograms showing artificial data. Left: mean = 1e-6; right: mean = 3. Standard deviation $\sigma = 10$ in both cases

II). The same standard deviation ($\sigma = 10$) is used in both cases. Histograms of the data are shown in Fig. 5.7.

In both examples, a one-sided test is performed as defined in (5.4) with the following parameters:

1. significance level: $\alpha = 0.05$
2. power: $1 - \beta = 0.8$
3. relevant difference: $\Delta = 2.5$
4. between-sample standard deviation: $\sigma = 10$.

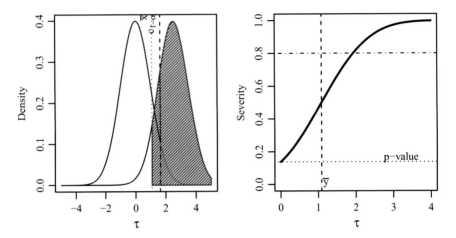

Fig. 5.8 Severity of not rejecting the null for a target relevant difference $\Delta = 2.5$. Right: Severity of not rejecting H_0 as a function of τ

Example: Data set I: Severity of not rejecting the null hypothesis

First, using data set I (100 samples from a $\mathcal{N}(\mu_1, \sigma^2)$ distributed random variable, with $\mu_1 = 1e - 6$ and $\sigma = 10$), the severity of not rejecting the null hypothesis is analyzed. Assume, the value $\bar{y} = 1.0889$ was observed, i.e., a statistically not significant difference (p-value $0.1381046 > 0.05$) is observed. But it would be a mistake to conclude from this result that the size of the difference is zero.

Figure 5.8 illustrates this situation by applying the concept of severity. The right panel in Fig. 5.8 provides a graphic depiction of answers to the following question for different values of τ: if the actual difference is at least τ, what is the probability of the observed estimate being higher than the actually observed value of $\bar{y} = 1.089$?

The greater this probability, the stronger the observed evidence is against that particular τ value. For two numbers, the answer is already known: for $\tau = 0$, namely 0.14, which is the p-value, and for $\tau = \bar{y}$, which is 50%. The p-value indicates that the null hypothesis of "no difference" cannot be rejected for $\alpha = 0.05$.

Because of the high variance in the data, the histogram is relatively broad (see the left panel in Fig. 5.7). This is now directly reflected in the assessment of other possible τ values (other initial hypotheses for a difference). For example, the severity of the evidence only crosses the threshold of 80% at a τ-value of approximately 2. This can be seen on the vertical axis in the right panel in Fig. 5.8. Therefore, if the actual difference was at least 2, then there would be a probability of 80% of estimating a value higher than the observed value 1.09. Even if the null hypothesis is not rejected, it cannot be concluded that the magnitude of the difference is zero. With high severity (80%), it can be concluded that the differences larger than 2 are unlikely. The right panel in Fig. 5.8 shows the severity of evidence (vertical axis) for all initial hypotheses (horizontal axis).

Table 5.2 Data set I: result analysis

p-value	Decision	Power	Cohen's d	Hedge's g	Severity
0.14	H0 not rejected	0.8037649	0.1212286	0.1212286	$\Delta \geq 2$ are well supported

The result statistic is presented in Table 5.2. The effect size suggests that the difference is of a smaller magnitude.

Example: Data set II: Severity of rejecting the null hypothesis

Data set II, with $\mu_2 = 3$ and $\sigma = 10$, is used to analyze the severity of rejecting the null hypothesis, i.e., the statistically significant estimate of $\bar{y} = 2.62$, resulting in the null hypothesis (of "there is no difference") being rejected, is considered.

Asserting this is evidence for a difference of exactly $\bar{y} = 2.62$ is not justified. Besides the null hypothesis, no further hypotheses were tested, e.g., "is the difference exactly 2.62?" or "is the difference at least 2.62?". Statistically, it was shown that there is a very low probability that there is no positive difference. So the evidence strongly ("severely") argues against the lack of an effect.

In the following, the test result is used to evaluate further hypotheses, e.g., that the difference is "not higher than at most τ," where τ represents a possible difference of, say, $\bar{y} = 4$ or $\bar{y} = 5$. The central question in this context is: How strongly does the experimental evidence speak against such an alternative null hypothesis, i.e., a difference of at most τ? This situation is comparable to the question of whether the null hypothesis can be rejected with sufficient certainty. This question can only be answered with a probability of error that can be estimated.

The following results were inspired by Bönisch and Inderst (2020), who present a similar discussion in the context of "damage estimation". For $\tau = 0$, the probability that the observed estimate is less than the observed value $\bar{y} = 2.62$ is $1 - p = 99.56$ %. Applying these results to other hypotheses about the value of τ leads to results shown in Fig. 5.9: For example, if $\tau = 1.75$, the probability of observing a value smaller than $\bar{y}$ is 80.84%. For $\tau = 2.5$, the probability would decrease to 54.85%. Consider—similar to the power value of 0.8—an 80% threshold as a minimum requirement for severity, then an estimate of $\bar{y} = 2.62$ that there is a sufficient severity against a difference up to $\tau = 1.75$ is obtained (Table 5.3).

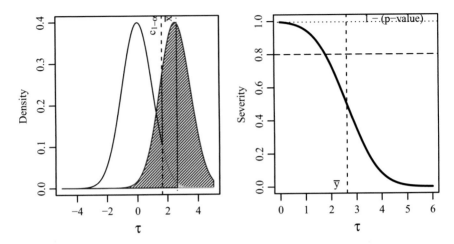

Fig. 5.9 Severity of rejecting H_0 (red), power (blue), and error (gray) for a target relevant difference $\Delta = 2.5$. Right: Severity of rejecting H_0 as a function of τ

Table 5.3 Data set II: result analysis

p-value	Decision	Power	Cohen's d	Hedge's g	Severity
0	H0 rejected	0.8037649	0.2737213	0.2737213	$\Delta \leq 2$ are well supported

5.7.4 Discussion of the 80% Threshold

Although a threshold of 80% was used in Fig. 5.9, it remains unclear at which threshold the level should be set. Demanding a severity of 90% has the consequence that even the assumption of a difference of at least 1.9 is not supported. Given that severity should also take into account domain-specific knowledge, a general value cannot be recommended. Visualizations such as Fig. 5.9 can help to get objective and rational results.

5.7.5 A Comment on the Normality Assumption

We discussed the extended classical hypothesis testing mechanism with Mayo's error statistics. Central tool in error statistics is severity, which allows a post-data analysis.

Severity can be applied to inferential statistics, no matter what the underlying distribution is Spanos (1999). We have discussed the normal distribution, because we will apply severity analysis to benchmarking (Sect. 5.8). In this context, the normality assumption holds, because most of the examples in this chapter use 50 or even more samples. The normality assumption is sometimes misunderstood: it does not require the population to resemble a normal distribution. It requires the sampling distribution of the mean difference should be approximately normal. In most cases, the central limit theorem will impart normality to the (hypothetical) distribution. This happens even to moderate n values, when the underlying population is not extremely asymmetric, e.g., caused by extreme outliers.

5.8 Severity: Application in Benchmarking

Now that we have the statistical tools available, i.e., power analysis plus error statistics (severity), we can evaluate their adequacy for scenarios in algorithm benchmarking.

The following experiments demonstrate how to perform a comparison of two algorithms. Our goal is not to provide a full comparison of many algorithms on many problems, i.e., MAMP, but to highlight important insights gained by severity. Therefore, two algorithms and three optimization problems were chosen. To cover the most important scenarios, three independent MASP studies will be performed. Each study compares two algorithms, say $\mathcal{A}$ and $\mathcal{B}$, on one problem instance.

The function `makeMoreFunList` from the R package `SPOTMisc` generates a list of functions presented in More et al. (1981), which is one of the most cited benchmark suites in optimization with more than 2000 citations. This list can be passed to the `runOptim` function, which performs the optimization. `runOptim` uses the arguments from Table 5.4.

We will compare the optimization methods CG and NM on the Rosenbrock, the Freudenstein and Roth, and Powell's Badly Scaled test function that were defined in More et al. (1981).

Table 5.4 `runOptim` arguments

Parameter	Description	Default value
fl	Function list	
method	The method used by `optim`: "Nelder-Mead", "BFGS", "CG", "L-BFGS-B", "SANN", or "Brent".	"Nelder-Mead"
n	Repeats. If $n > 1$, different start points (randomized) will be used	2
k	Subset of benchmark functions	All implemented functions
verbosity	Level of information to be shown	0

5.8.1 Experiment I: Rosenbrock

5.8.1.1 Pre-experimental Planning

In our first experiment, we will use the Rosenbrock function; see Eq. 5.6. This is the first function in More et al. (1981)s study, so we will pass the argument $k = 1$ to the `runopt()` function. To estimate the number of function evaluations, a few pre-experimental runs of the algorithms are performed. These pre-experimental runs are also necessary for testing numerical instabilities, expected behavior, and correct implementations. In our case, $n = 20$ pre-experimental runs were performed.

```
library("SPOT")
set.seed(1)
k <- 1 # More function no. 1
n0 <- 20 # Pre-experimental runs
moreF1 <- makeMoreFunList()
resCG0 <- runOptim(
  f1 = moreF1,
  method = "CG",
  n = n0,
  k = k
)
resNM0 <- runOptim(
  f1 = moreF1,
  method = "Nelder-Mead",
  n = n0,
  k = k
)
```

A data.frame with 20 observations is available for each algorithm, e.g., for CG:

```
str(resCG0)

## 'data.frame':     20 obs. of  3 variables:
##  $ f: num  1 1 1 1 1 1 1 1 1 1 ...
##  $ r: num  1 2 3 4 5 6 7 8 9 10 ...
##  $ y: num  0.10644 0.0686 0.00577 0.08434 3.65499 ...
```

Looking at the summary of the results is strongly recommended. R's `summary` is the first choice.

```
summary(resCG0$y)

##      Min.  1st Qu.   Median     Mean  3rd Qu.      Max.
## 0.000939 0.068510 0.080510 0.549287 0.119892 3.654986
```

```
summary(resNM0$y)

##       Min.    1st Qu.    Median      Mean   3rd Qu.      Max.
## 1.900e-08  5.740e-07 1.439e-06 1.622e-04 3.233e-06 3.206e-03
```

The summaries indicate that NM is superior and that CG has some outliers. A graphical inspection is shown in Fig. 5.10. Taking care of extreme outliers is recommended in further analysis.

We are interested in the mean difference in the methods' performances. The pre-experimental runs indicate that the difference is $\bar{y} = 0.55$. Because this value is positive, we can assume that method NM is superior. The standard deviation is $s_d = 1.14$. Based on Eq. 5.14, and with $\alpha = 0.05$, $\beta = 0.2$, and $\Delta = 0.5$, we can determine the number of runs for the full experiment with the `getSampleSize()` function.

For a relevant difference of 0.5, approximately 65 completing runs per algorithm are required. Figure 5.11 illustrates the situation for various Δ and three n values.

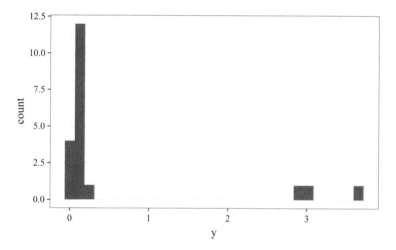

Fig. 5.10 Results from CG on Rosenbrock. Histogram to inspect outliers

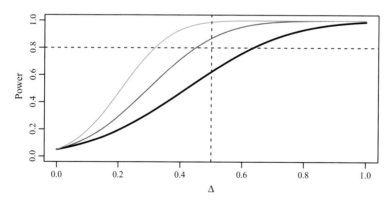

Fig. 5.11 Rosenbrock (function 1). Power as a function of the relevant difference Δ for a two-parallel-group experiment (black = 40, red = 80, and green = 160 runs). If the relevant difference is 0.5, n = 160 runs per algorithm are needed for 80% power

Although we do not know any "true" relevant difference for artificial (dimension less) test functions, we consider the distance $\Delta = 0.5$ as relevant and, to play safe, choose $n = 80$ algorithm runs for the full experiment.

5.8.1.2   Performing the Experiments on Rosenbrock

The full experiments can be conducted as follows. The 20 results from the pre-experimental runs will be "recycled", only 60 additional runs must be performed. How to combine existing results with new ones was discussed in Sect. 4.5.3. The corresponding code is similar to the code that was used for the pre-experimental experiments in Sect. 5.8.1.1.

```
summary(resCG$y)
```

```
##      Min.  1st Qu.    Median      Mean  3rd Qu.      Max.
## 0.000053 0.037231 0.079467 0.681190 0.107427 4.332730
```

```
summary(resNM$y)
```

```
##      Min.  1st Qu.    Median      Mean  3rd Qu.      Max.
## 4.000e-09 1.430e-07 8.260e-07 8.207e-05 2.686e-06 3.206e-03
```

Figure 5.12 shows a histogram of the results.
The numerical summary of these results is

```
##      Min.  1st Qu.    Median      Mean  3rd Qu.      Max.
## 0.000052 0.037222 0.079464 0.681108 0.107421 4.332730
```

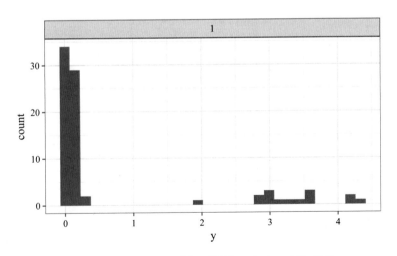

Fig. 5.12 Rosenbrock: Difference between CG and NM results (y = CG - NM)

Table 5.5 Experiment I: result analysis

p-value	Decision	Power	Cohen's d	Hedge's g	Severity
0	H0 rejected	0.961397	0.7348167	0.7313231	$\Delta \le 0.5$ are well supported

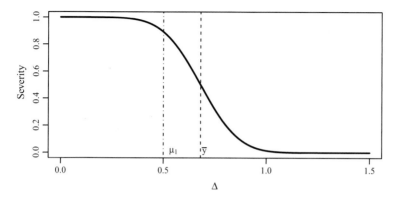

Fig. 5.13 Rosenbrock. Severity for rejecting H_0. Given data from the experiment, claiming that the true difference is as large or larger than 1.0 has a very low severity, whereas differences as large as 0.5 are well supported by severity

The sample mean of the differences is $\bar{y} = 0.68$. Obviously, NM is superior and there is a difference in performance. The question remains: how large is this difference? To answer this question, we will analyze results from these runs with severity.

The summary result statistic is presented in Table 5.5. The effect size suggests that the difference is of medium magnitude. The corresponding severity plot is shown in Fig. 5.13.

5.8.1.3 Discussion

Results indicate that the NM method is superior. Beyond the classical analysis based on EDA tools and hypothesis tests, severity allows further conclusions: It shows that performance differences smaller than 0.5 are well supported. Although the situation is clear, the final choice is up to the experimenter. They might include additional criteria such as run time, costs, and robustness in their final decision. And last but not least: The question of whether a difference of 0.5 is of practical relevance is highly dependent on external factors.

5.8.2 Experiment II: Freudenstein-Roth

The two-dimensional Freudenstein and Roth Test Function (Freudenstein and Roth 1963), which is number $k = 2$ in More et al. (1981)s list, will be considered next. The function is defined as

$$f(x_1, x_2) = (x_1 - 13 + ((5 - x_2)x_2 - 2)x_2)^2 + (x_1 - 29 + ((1 + x_2)x_2 - 14)x_2)^2.$$

5.8.2.1 Pre-experimental Planning: Freudenstein and Roth

Similar to the study of the Rosenbrock function, 20 pre-experimental runs are performed. We take a look at the individual results.

```
summary(resCG0$y)

##    Min. 1st Qu.  Median    Mean 3rd Qu.    Max.
## 0.3928 19.6811 77.9517 57.7497 81.2694 86.3797

summary(resNM0$y)

##    Min. 1st Qu.  Median    Mean 3rd Qu.    Max.
##   48.98   48.98   48.98   48.98   48.98   48.98
```

The summaries indicate that NM is not able to find improved values. A floor effect occurred (Bartz-Beielstein 2006). The experiment is too difficult for NM. No further experiments will be performed, because NM is not able to find improvements. A re-parametrization of the NM (via hyperparameter tuning) is recommended, before additional experiments are performed.

Although CG appears to be superior and can find values as small as 0.3928, it has problems with outliers as can be seen in Fig. 5.14.

5.8.2.2 Discussion

An additional, experimental performance analysis (that focuses on the mean) is not recommended in this case, because the result is clear: CG outperforms NM.

5.8.3 Experiment III: Powell's Badly Scaled Test Function

Powell's two-dimensional Badly Scaled Test function, which is number $k = 3$ in More et al. (1981)s list, will be considered next.

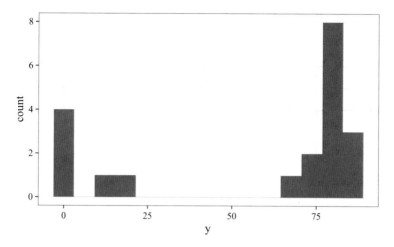

Fig. 5.14 Results from method CG on Freudenstein Roth. Histogram to inspect outliers

The function is defined as

$$f(x_1, x_2) = f_1^2 + f_2^2$$

with

$$f_1 = 1e4x_1x_2 - 1 \quad \text{and} \quad f_2 = \exp(-x_1) + \exp(-x_2) - 1.0001.$$

5.8.3.1 Pre-experimental Planning: Powell's Badly Scaled Test Function

First, we take a look at the individual results. The summaries do not clearly indicate which algorithm is superior. A graphical inspection is shown in Fig. 5.15. Both methods are able to find improvements, but both are affected by outliers. The pre-experimental runs indicate that the difference is $\bar{y} = -0.21$. Because this value is negative, we will continue the analysis under the assumption (hypothesis) that method CG is superior.

We are interested in the mean difference in the algorithms' performances.

The standard deviation is $s_d = 1.5$. Based on Eq. 5.14, and with $\alpha = 0.05$, $\beta = 0.2$, and $\Delta = 0.5$, we can determine the number of runs for the full experiment.

For a relevant difference of 0.5, approximately 112 completing runs per algorithm are required. Figure 5.16 illustrates the situation for various Δ and three n values.

Although we do not know any "true" relevant difference for artificial (dimension less) test functions, we consider a distance $\Delta = 0.5$ as relevant and choose $n = 120$ algorithm runs for the full experiment.

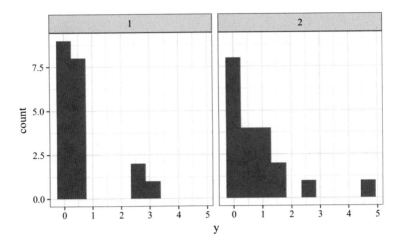

Fig. 5.15 Results from CG and NM on Powell's badly scaled test function. Histograms to inspect outliers

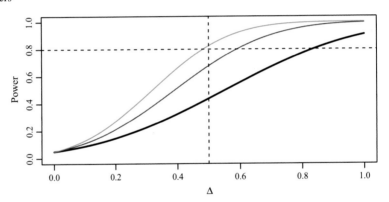

Fig. 5.16 Powell's badly scaled test function: Power as a function of the relevant difference Δ for a two-parallel-group experiment (black = 40, red =80, and green = 120 runs). If the relevant difference is 0.5, n = 120 runs per algorithm are needed for 80% power

5.8.3.2 Performing the Experiments: Powell's Badly Scaled Test Function

The full experiments can be conducted as follows. Results from the pre-experimental runs will be "recycled", only 100 additional runs must be performed.

A graphical inspection is shown in Fig. 5.17, which shows a histogram of the results. As expected, both algorithms are able to find improvements, but are affected by outliers.

Figure 5.18 shows a histogram of the differences. The numerical summary of these results is

```
## [1] "CG"
##       Min.   1st Qu.    Median       Mean   3rd Qu.        Max.
##   0.003772  0.094777  0.323066  0.998882  0.710185 22.565737
## [1] "NM"
##       Min.   1st Qu.   Median       Mean  3rd Qu.       Max.
## 0.000026  0.110220  0.329906  0.757107 1.040480 8.932574
## [1] "diff"
##       Min.   1st Qu.   Median       Mean  3rd Qu.       Max.
## -8.81704  -0.25991  -0.02496   0.24177  0.36554 21.58666
```

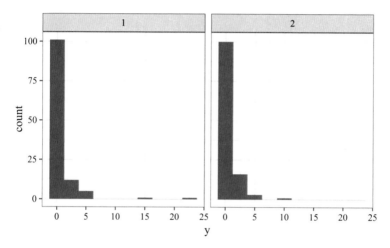

Fig. 5.17 Results from CG and NM on Powell's badly scaled test function (n = 120). Histograms to inspect outliers

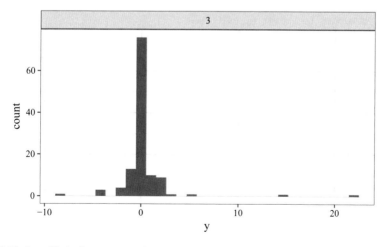

Fig. 5.18 Powell's badly scaled test function: Difference between CG and NM results (y = CG–NM)

Table 5.6 Experiment III: Result Analysis

p-value	Decision	Power	Cohen's d	Hedge's g	Severity
0.17	H0 not rejected	0.6274005	0.1184404	0.1180668	$\Delta \geq 0.5$ are well supported

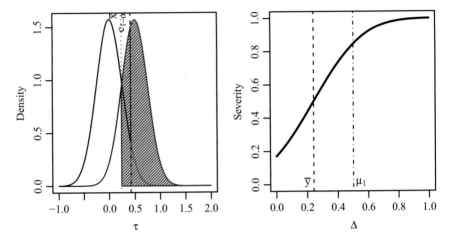

Fig. 5.19 Powell's badly scaled test function: Left: Severity of not rejecting H_0 (red), power (blue) for a target relevant difference of $\Delta = 0.5$. Right: Severity of not rejecting H_0 as a function of Δ

The summary result statistic is presented in Table 5.6. The effect size suggests that the difference is of a smaller magnitude. For the chosen $\Delta = 0.5$, the severity value is at 0.85 and thus it strongly supports the decision of not rejecting the H_0.

5.8.3.3 Discussion

Results from these runs can be analyzed using severity. The sample mean of the differences is $\bar{y} = 0.24$, so method NM might be superior. However, the median is negative. It is not obvious, which method is superior. The corresponding severity plot is shown in Fig. 5.19.

5.9 Summary and Discussion

Simply proving that there is a difference between the performance of the methods, e.g., by performing a one-sided test, is in many situations not sufficient: one needs to show that this difference is relevant. Severity was introduced as one way to tackle this problem.

A research question is necessary, e.g., if we are facing a real-world problem that has similar structural properties (discovered by landscape analysis; see Mersmann et al. 2011) as the artificial test function. Then, it might be interesting to see whether a gradient-based method (CG) is superior compared to a gradient-free method (NM). Even when theoretical results are available, they should be validated (numerical instabilities, dependencies on starting points, etc.).

Finally, at the end of this chapter, we may ask: *Why severity?* An optimization algorithm, e.g., $\mathcal{A}^+$, has achieved a high success rate with a test problem: the optimum can be determined in 96.3% of the cases. Consider the following situations:

- Let us first assume that an algorithm, say $\mathcal{A}^-$, which has no domain knowledge, only achieves such a high success rate as $\mathcal{A}^+$ in very rare exceptional cases. Is this score a good indication that $\mathcal{A}^+$ is well suited to solve this problem? In this case, based on the test results of $\mathcal{A}^+$ and $\mathcal{A}^-$, the conclusion would be justified.
- Next, suppose that Algorithm $\mathcal{A}^-$, which does not use domain knowledge, would have no problem having a score of up to 96%. Again, we can ask the same question: is this 96.3% score good evidence that $\mathcal{A}^+$ is well suited for this test problem? Based on information about the results of $\mathcal{A}^+$ and $\mathcal{A}^-$, in this case the conclusion would rather not be justified.

The severity provides a meta-statistical concept to identify these effects, which are also known in the literature as floor and ceiling effects (Cohen 1995).

Part II
Applications

Chapter 6
Hyperparameter Tuning and Optimization Applications

Thomas Bartz-Beielstein

Abstract This chapter reflects on advantages and sense of use of Hyperparameter Tuning (HPT) and its disadvantages. In particular it shows how important it is, to keep the human in the loop, even if HPT works perfectly. The chapter presents a collection of HPT studies. First, HPT applications in Machine Learning (ML) and Deep Learning (DL) are described. A special focus lies on automated ML, neural architecture search, and combined approaches. HPT software is presented. Finally, model based approaches, especially applications with Sequential Parameter Optimization Toolbox (SPOT) are discussed.

6.1 Surrogate Optimization

Starting in the 1960s, Response Surface Methodology (RSM) and related Design of Experiments (DOE) methods were transferred from the engineering domain (e.g., from physics, agriculture, chemistry, and aerospace) to computer science (Montgomery 2017). With the increasing computational power, computer simulations, scientific computing and computational statistics gained importance (Gentle et al. 2004; Strang 2007). Kleijnen (1987) summarizes these ideas and methods in a very comprehensible manner for simulation practitioners. After computer simulations replaced expensive lab experiments, these computer simulations themselves were substituted by even cheaper computer models: surrogate models or in short, *surrogates*, that imitate complex numerical calculations, were developed. Kriging surrogates (or Gaussian Processs (GPs) aka Bayesian Optimization (BO)), that gleaned ideas from computer experiments in geostatistics, enjoy wide applicability, especially in domains where predictions are required. Today, GP models are used as powerful predictors for all sorts of applications in engineering and ML. GP methods replaced classical regression methods in many domains. Santner et al. (2003), Forrester et al. (2008b), and Gramacy (2020) wrote groundbreaking works in this field. In global optimization, Efficient Global Optimization (EGO) became a very popular approach

T. Bartz-Beielstein (✉)
Institute for Data Science, Engineering and Analytics, TH Köln, Cologne, Germany
e-mail: thomas.bartz-beielstein@th-koeln.de

© The Author(s) 2023
E. Bartz et al. (eds.), *Hyperparameter Tuning for Machine and Deep Learning with R*,
https://doi.org/10.1007/978-981-19-5170-1_6

(Schonlau 1997; Jones et al. 1998). Emmerich (2005) showed that surrogates can significantly accelerate evolutionary multi-objective optimization algorithms in the presence of time consuming evaluations.

SPOT was one of the first approaches that combined classical regression, surrogate optimization (Kriging), and, especially for non-continuous variables, decision trees, for the optimization of algorithm (hyper-)parameters (Bartz-Beielstein et al. 2004, 2005). Applied in hyperparameter optimization in Evolutionary Computation (EC), see, e.g., Lobo et al. (2007), the collection *Experimental Methods for the Analysis of Optimization Algorithms* includes important publications in this field (Bartz-Beielstein et al., 2010). For example, the following contributions paved the way for important developments in HPT:

- Ridge and Kudenko (2010) describe the classical DOE approach, e.g., how to set up experimental designs, for algorithm benchmarking.
- The contribution *Sequential Model-Based Optimization* by Hutter et al. (2010b) laid the foundation for surrogate optimization in ML and resulted in software tools such as Sequential Model-Based Optimization for General Algorithm Configuration (SMAC) (Hutter et al. 2010a).
- Iterative Racing (IRACE), which is a generalization of the Iterated F-race method, is another popular tool for the automatic configuration of optimization algorithms (Birattari et al. 2009).

Surrogate optimization is the *de facto* standard for complex optimization problems, especially for continuous variables. Bartz-Beielstein and Zaefferer (2017) presented methods for continuous and discrete optimization based on Kriging. While surrogate models are well-established in the continuous optimization domain, they are less frequently applied to more complex search spaces with discrete or combinatorial solution representations (Zaefferer et al. 2014). Zaefferer (2018) Ph.D. thesis fills this gap, showing how surrogate models like Kriging can be extended to arbitrary problem types, if measures of similarity between candidate solutions are available. Today, surrogate optimization is applied in the engineering domain as well as in computer science, e.g., for HPT.

Example: Mixed-Discrete Problems

Many real-world optimization problems consider the optimization of ordinal integers, categorical integers, binary variables, permutations, strings, trees, or graphs structures in general. These real-world problems pose complex search spaces which require a deep understanding of the underlying solution representations.

Some of them, for example integers, are more suitable to be treated by classic optimization algorithms. Others, such as trees, have to be handled by specifically developed optimization algorithms. In general, solving these kinds of problems usually necessitates a significant number of objective function evaluations. However, in many engineering problems, a single evaluation is based on either on experimental or numerical analysis. This causes significant costs with respect to time or resources.

Surrogate Model Based Optimization (SMBO) aims to handle the complex variable structures and the limited budget simultaneously. Sequential Parameter Optimization (SPO) pursues the identification of global optima taking advantage of a budget allocation process that maximizes the information gain in promising regions. Gentile et al. (2021) presented an efficient method to face mixed-discrete optimization problems using surrogates.

Example: Alzheimer's Disease

Bloch and Friedrich (2021) used ML for early detection of Alzheimer's disease especially based on magnetic resonance imaging. The authors use BO to time-efficiently find good hyperparameters for Random Forest (RF) and Extreme Gradient Boosting (XGBoost) models, which are based on four and seven hyperparameters and promise good classification results. Those models are applied to distinguish if mild cognitive impaired subjects from the Alzheimer's disease neuroimaging initiative data set will prospectively convert to Alzheimer's disease.

The results showed comparable Cross Validation (CV) classification accuracies for models trained using BO and grid-search, whereas BO has been less time-consuming. Similar to the approaches presented in this book (and in many other BO studies), the initial combinations for BO were set using Latin Hypercube Design (LHD) and via random initialization. Furthermore, many models trained using BO achieved better classification results for the independent test data set than the model based on the grid-search. The best model was an XGBoost model trained with BO.

Example: Elevator Simulation and Optimization

Modern elevator systems are controlled by the elevator group controllers that assign moving and stopping policies to the elevator cars. Designing an adequate Elevator Group Control (EGC) policy is challenging for a number of reasons, one of them being conflicting optimization objectives. Vodopija et al. (2022) address this task by formulating a corresponding constrained multiobjective optimization problem, and, in contrast to most studies in this domain, approach it using true multiobjective optimization methods capable of finding approximations for Pareto-optimal solutions.

Specifically, they apply five multiobjective optimization algorithms with default constraint handling techniques and demonstrate their performance in optimizing EGC for nine elevator systems of various complexity. SPO was used to tune the algorithm parameters. The experimental results confirm the scalability of the proposed methodology and suggest that NSGA-II equipped with the constrained-domination principle is the best performing algorithm on the test EGC systems. The proposed problem formulation and methodology allow for better understanding of the EGC design problem and provide insightful information to the stakeholders involved in deciding on elevator system configurations and control policies.

Example: Cyber-physical Production Systems

Bunte et al. (2019) developed a cognitive architecture for Artificial Intelligence (AI) in Cyber-physical Production Systemss (CPPSs). The goal of this architecture is to reduce the implementation effort of AI algorithms in CPPSs. Declarative user goals and the provided algorithm-knowledge base allow the dynamic pipeline orchestration and configuration. A big data platform instantiates the pipelines and monitors the CPPSs performance for further evaluation through the cognitive module. Thus, the cognitive module is able to select feasible and robust configurations for process pipelines in varying use cases. Furthermore, it automatically adapts the models and algorithms based on model quality and resource consumption. The cognitive module also instantiates additional pipelines to evaluate algorithms from different classes on test functions.

Example: Resource Planning in Hospitals

Pandemics pose a serious challenge to health-care institutions. To support the resource planning of health authorities from the Cologne region, BaBSim.Hospital, a tool for capacity planning based on discrete event simulation, was created Bartz-Beielstein et al. (2021a). The predictive quality of the simulation is determined by 29 parameters with reasonable default values obtained in discussions with medical professionals. Bartz-Beielstein et al. (2021b) aimed to investigate and optimize these parameters to improve BaBSim.Hospital using a surrogate optimization approach and an in-depth sensitivity analysis.

Because SMBO is the default method in optimization via simulation, there are many more examples from the application domain, e.g., Waibel et al. (2019) present methods for selecting tuned hyper-parameters of search heuristics for computationally expensive simulation-based optimization problems.

6.2 Hyperparameter Tuning in Machine and Deep Learning

In contrast to HPT in optimization, where the objective function with related input parameters is clearly specified for the tuner, the situation in ML is more complex. As illustrated in Fig. 2.2, the tuner is confronted with several loss functions, metrics, and data sets. As discussed in Sect. 2.3, there is no clear answer to this problem.

Furthermore, the situation in ML is more challenging than in optimization, because ML methods develop an increasing complexity. Although for specific problems, especially when domain knowledge is available, methods such as Support Vector Machine (SVM) or Elastic Net (EN) cannot be beaten by more complex methods—especially under tight time and computational constraints. In these well specified settings, hand-crafted SVM kernel methods cannot be beaten by complex methods. Therefore, in some well-defined domains, SVMs and related shallow methods can still be considered as efficient methods.

With increasing computational power and memory, more and more complex methods gain popularity. Some ML standard methods are cheap to evaluate, e.g., Decision Tree (DT), model complexity and performance increase: RF replaced simple trees and even more sophisticated methods such as XGBoost are considered state of the art.

The situation is getting worse when Deep Neural Networks (DNNs) are included: there is no limit for model complexity. As a consequence, HPT is developing very quickly to catch up with this exploding model complexity. New branches and extensions of existing HPT branches were proposed, e.g., Combined Algorithm Selection and Hyperparameter optimization (CASH), Neural Architecture Search (NAS), Automated Hyperparameter and Architecture Search (AutoHAS), and further "Auto-*" approaches (Thornton et al. 2013; Dong et al. 2020). The jungle of new ML and DL is accompanied by a plethora of HPT approaches and related software tools. Hutter et al. (2019) presents an overview of Automated Machine Learning (AutoML).

Although DL has been part of AI for a long time—first ideas go back to the 1940s—its break through happened in the early 2010s (McCulloch and Pitts 1943; Krizhevsky et al. 2012). Since 2012, Convolutional Neural Networks (CNNs) are the dominating approach in computer vision and image classification. In parallel, DL was adopted in several other domains, e.g., Natural Language Processing (NLP). DL methods outperformed standard ML methods such as SVMs in a wide range of applications (Chollet and Allaire 2018). Although finding good hyperparameters for shallow methods like SVM can be a challenging task, DL methods increased the difficulty significantly, because they explode the dimensionality of the hyperparameter space, Λ.

Therefore, it is worth looking at HPT strategies that were developed for DL. For example, Snoek et al. (2012) used the Canadian Institute for Advanced Research, 10 classes (CIFAR-10) data set, which consists of 60,000 32×32 color images in ten classes, for optimizing the hyperparameters of a CNNs. Bergstra et al. (2013) proposed a meta-modeling approach to support automated Hyperparameter Optimization (HPO), with the goal of providing practical tools that replace hand-tuning. They optimized a three-layer CNN. Eggensperger et al. (2013) collected a library of HPO benchmarks and evaluated three BO methods. Zoph et al. (2017) studied a new paradigm of designing CNN architectures and describe a scalable method to optimize these architectures on a data set of interest, for instance, the ImageNet classification data set.

The following example describes a typical approach of HPT in DL.

Example: Robust and Efficient Hyperparameter Optimization in DL

Falkner et al. (2018) optimized six hyperparameters that control the training procedure of a fully connected DNN (initial learning rate, batch size, dropout, exponential decay factor for learning rate) and the architecture (number of layers, units per layer)

Table 6.1 Elements of a typical HPT study: The hyperparameters and architecture choices for the fully connected networks as defined in Falkner et al. (2018)

Hyperparameter	Lower bound	Upper bound	Log-transform
Batch size	2^3	2^8	Yes
Dropout rate	0	0.5	No
Initial learning rate	$1e-6$	$1e-2$	Yes
Exponential decay factor	-0.185	0	No
# hidden layers	1	5	No
# units per layer	2^4	2^8	Yes

for six different data sets gathered from OpenML (Vanschoren et al. 2014), see Table 6.1.

The authors used a surrogate DNN as a substitute for training the networks directly. To build this surrogate, they sampled 10 000 random configurations for each data set, trained them for 50 epochs, and recorded the classification error after each epoch, and total training time. Two independent RF models were fitted to predict these two quantities as a function of the hyperparameter configuration used. Falkner et al. (2018) noted that Hyperband (HB) initially performed much better than the vanilla BO methods and achieved a roughly three-fold speedup over Random Search (RS).

Artificial toy functions were used in this study, and because BO does not work well on high-dimensional mixed continuous and categorical configuration spaces, they used a simple counting-ones problem to analyze this issue.

Tip: Handling mixed continuous, categorical, and combinatorial configuration spaces

Zaefferer et al. (2014) discussed these topics in great detail. How to implement BO for discrete (and continuous) optimization problems was analyzed in the seminal paper by Bartz-Beielstein and Zaefferer (2017). Furthermore, Zaefferer (2018) provides an in-depth treatment of this topic. In practice, SPOT can handle categorical and mixed variables as discussed in Sect. 4.5. Combinatorial problems, such as the optimization of permutations, strings, or graphs, can be treated by the R package CEGO (Zaefferer 2021).

Kedziora et al. (2020) analyzed what constitutes these systems and survey developments in HPO, e.g., multi-component models, Neural Network (NN) architecture search, automated feature engineering, meta-learning, multi-level ensembling, multiobjective evaluation, flexible user involvement, and principles of generalization, to name only a few.

Wistuba et al. (2019) described how complex DL architectures can be seen as combinations of a few elements, so-called *cells*, that are repeated to build the complete network. Zoph and Le (2016) were the first who proposed a cell-based approach, i.e., choices made about a NN architecture are the set of meta-operations and their arrangement within the cell. Another interesting example is function-preserving morphisms implemented by the Auto-Keras package to effectively traverse potential networks Jin et al. (2019).

NAS is discussed in (NAS Elsken et al. (2019)). Mazzawi et al. (2019) introduced a NAS framework to improve keyword spotting and spoken language identification models. Lindauer and Hutter (2020) describe AutoML for NAS.

Because optimizers can affect the DNN performance significantly, several tuning studies devoted to optimizers were published during the last years: Schneider et al. (2019) introduced a benchmarking framework called Deep Learning Optimizer Benchmark Suite (DeepOBS), which includes a wide range of realistic DL problems together with standardized procedures for evaluating optimizers. Schmidt et al. (2020) performed an extensive, standardized benchmark of fifteen particularly popular DL optimizers.

Menghani (2021) presented a survey of the core areas of efficiency in DL, e.g., spanning modeling techniques, infrastructure, and hardware accompanied by an experiment-based guide along with code for practitioners to optimize their model training and deployment.

Tunability, (see Definition 2.26) is an interesting concept that should be mentioned in the context of HPT (Probst et al. 2019a). The hope is that identifying *tunable* hyperparameters, i.e., ones that model performance is particularly sensitive to, will allow other settings to be ignored and results in a reduced hyperparameter search space, Λ. Unfortunately, tunability strongly depends on the choice of the data set, $(\mathcal{X}, \mathcal{Y})$, which makes a generalization of results very difficult.

Bischl et al. (2021a) provide an overview about HPO.

6.3 HPT Software Tools

The field of HPT software tools is under rapid development. Besides SPOT, which is discussed in this book, several other hyperparameter optimization software packages were developed. We will list packages that show a certain continuity and that hopefully will still be available in the near future.

The `irace` package implements the Iterated Race method, which is a generalization of the Iterated F-race method for the automatic configuration of optimization algorithms. Hyperparameters are tuned by finding the most appropriate settings for a given set of instances of an optimization problem. It builds upon the race package by Birattari et al. (2009) and it is implemented in R (López-Ibáñez et al. 2016).

The Iterative Optimization Heuristics profiler (IOHprofiler) is a benchmarking and profiling tool for optimization heuristics, composed of two main components (Doerr et al. 2018): The Iterative Optimization Heuristics analyzer (IOHanalyzer) provides an interactive environment to evaluate algorithms' performance by various criteria, e.g., by means of the distribution on the fixed-target running time and the fixed-budget function values (Wang et al. 2022). The experimental platform, Iterative Optimization Heuristics experimenter (IOHexperimenter), is designed to ease the generation of performance data. Its logging functionalities allow to track the evolution of algorithm parameters, making the tool particularly useful for the analysis, comparison, and design of algorithms with (self-)adaptive hyperparameters. Balaprakash et al. (2018) presented DeepHyper, a PYTHON package that provides a common interface for the implementation and study of scalable hyperparameter search methods. Karmanov et al. (2018) created a *"Rosetta Stone"* of DL frameworks to allow data scientists to easily leverage their expertise from one framework to another. They provided a common setup for comparisons across GPUs (potentially CUDA versions and precision) and for comparisons across languages (Python, Julia, R). O'Malley et al. (2019) presented *Keras tuner*, a hyperparameter tuner for Keras with TensorFlow (TF) 2.0. Available tuners are RS and Hyperband. Mendoza et al. (2019) introduced *Auto-Net*, a system that automatically configures NN with SMAC by following the same AutoML approach as Auto-WEKA and Auto-sklearn. Zimmer et al. (2020) developed Auto-PyTorch, a framework for Automated Deep Learning (AutoDL) that uses Bayesian Optimization HyperBand (BOHB) as a back-end to optimize the full DL pipeline, including data pre-processing, network training techniques, and regularization methods. Mazzawi and Gonzalvo (2021) presented Google's Model Search, which is an open-source platform for finding optimal ML models based on TF. It does not focus on a specific domain.

Unfortunately, many of these software tools are results from research projects that are funded for a limited time span. When the project ends (and the developers successfully completed their Ph.D.s) the software package will not be maintained anymore. Despite the dynamics and volatility in this area, we do not want to shy away from giving an overview of the available software tools. Table 6.2 presents this overview, which should be regarded as an incomplete snapshot, but not as the whole picture of this field.

6.4 Summary and Discussion

Due to increased computational power, algorithm and model complexity grow into new regions. It is more and more important to understand the working mechanisms of complex neural networks. Putting the pieces together, it becomes clear that

1. there is a need for hyperparameter tuning,
2. surrogate optimization is an efficient approach, it can accelerate the search, and
3. mixed variable types (continuous, discrete) make hyperparameter tuning more difficult. Especially dependencies between different hyperparameters produce new challenges.

Table 6.2 Overview: HPT and HPO approaches

Software	Application	Method	Publication
AutoPyTorch	Fully automated DL (AutoDL)	BOHB	Zimmer et al. (2020)
Auto-Sklearn	Automated ML toolkit	BO, meta-learning and ensemble construction	Feurer et al. (2020)
Auto-WEKA	Search for the right WEKA ML algorithm and optimizes its hyperparameters	BO	Kotthoff et al. (2017)
BOHB	Distributed HB	BO and bandit-based methods	Falkner et al. (2018)
CAVE	Report generation	EDA, parameter importance analysis	Biedenkapp et al. (2018)
DEHB	Black-box optimization	HB, DE	Awad et al. (2021)
Google's model search	Build on TF, architecture search	multiple trainers, a search algorithm, a transfer learning algorithm. Database to store ML and DL models	Mazzawi and Gonzalvo (2021)
Hyperopt	Python library for serial and parallel optimization, can handle real-valued, discrete, and conditional dimensions	RS and TPEs	Bergstra et al. (2013), Koehrsen (2018)
IOHprofiler, IOHanalyzer, IOHexperimenter	analyze and visualize the empirical performance of IOHs, interactive plotting, statistical evaluation, report generation	R packages `Shiny`, `Plotly`, `Rcpp`	Doerr et al. (2018), Wang et al. (2022)
irace	Heuristics, automatic configuration of optimization and decision algorithms, appropriate settings of an algorithm given a set of instances of a problem	iterated racing	López-Ibáñez et al. (2016)
keras tuner	Hyperparameter tuner for keras/TF	RS, HB	O'Malley et al. (2019)
mlmachine	Uses Hyperopt as a foundation for performing experiments	BO	Koehrsen (2018)
Optuna	Software framework for ML	TPE, RS, grid search, CMA-ES	Akiba et al. (2019)
Ray-Tune	PyTorch, XGBoost, MXNet, and Keras and other frameworks	Wrapper around open-source optimization libraries such as HyperOpt, SigOpt, Dragonfly, and Facebook Ax	Liaw et al. (2018)
SMAC	Tool for algorithm configuration	BO, racing mechanism	Lindauer et al. (2022)
SPOT	Surrogate optimization	Various surrogates and optimizers, BO, RSM	Bartz-Beielstein et al. (2017)

To conclude this chapter, we would like to mention relevant criticism of HPT. Some authors even claimed that extensive HPT is not necessary at all. For example, Erickson et al. (2020) introduced a framework (AutoGluon-Tabular) that "requires only a single line of PYTHON to train highly accurate machine learning models on an unprocessed tabular data set such as a CSV file". AutoGluon-Tabular ensembles several models and stacks them in multiple layers. The authors claim that AutoGluon-Tabular outperforms AutoML platforms such as TPOT, H2O, AutoWEKA, auto-sklearn, AutoGluon, and Google AutoML Tables.

A highly recommendable study was performed by Choi et al. (2019), who presented a taxonomy of first-order optimization methods. Furthermore, Choi et al. (2019) demonstrated the sensitivity of optimizer comparisons to the hyperparameter tuning protocol: "optimizer rankings can be changed easily by modifying the hyperparameter tuning protocol." These results raise serious questions about the practical relevance of conclusions drawn from certain ways of empirical comparisons. They also claimed that tuning protocols often differ between works studying NN optimizers and works concerned with training NNs to solve specific problems.

Yu, Sciuto, Jaggi, Musat, and Salzmann (Yang and Shami) claimed that the evaluated state-of-the-art NAS algorithms do not surpass RS by a significant margin, and even perform worse in the Recurrent Neural Network (RNN) search space.

Balaji and Allen (2018) reported a multitude of issues when attempting to execute automatic ML frameworks. For example, regarding the random process, the authors state that "one common failure is in large multi-class classification tasks in which one of the classes lies entirely on one side of the train test split".

Li and Talwalkar (2019) stated that (i) better baselines that accurately quantify the performance gains of NAS methods, (ii) ablation studies (to learn about the NN by removing parts of it) that isolate the impact of individual NAS components, and (iii) reproducible results that engender confidence and foster scientific progress are necessary.

Liu (2018) remarks that "for most existent AutoML works, regardless of the number of layers of the outer-loop algorithms, the configuration of the outermost layer is definitely done by human experts". Human experts are shifted to a higher level, and are still in the loop. The lack of insights in current AutoML systems (Drozdal et al. 2020) goes so far that some users even prefer manual tuning as they believe they can learn more from this process (Hasebrook et al. 2022).

Taking this criticism seriously, we can conclude that transparency and interpretability of *both* the ML / DL method *and* the HPT process are mandatory. This conclusion becomes very important in safety-critical applications, e.g., security-critical infrastructures (drinking water), in medicine, or automated driving.

But in general, we can conclude, that HPT is a valuable, in some situations an even mandatory tool for understanding ML and DL methods. And, last but not least: HPT tools can help to gain trust in AI systems.

Chapter 7
Hyperparameter Tuning in German Official Statistics

Florian Dumpert and Elena Schmidt

Abstract This chapter describes the special quality requirements placed on official statistics and builds a bridge to the tuning of hyperparameters in Machine Learning (ML). To carry out the latter optimally under consideration of constraints and to assess its quality is part of the tasks of the employees entrusted with this work. The chapter sheds special light on open questions and the need for further research.

7.1 Official Statistics

Official (federal) statistics in Germany (as in many other countries) have a special mandate: to provide government, parliament, interest groups, science, and the public with information on the most diverse areas of economic, social, and cultural life, the state, and the environment. More precisely, § 1 of the *Gesetz über die Statistik für Bundeszwecke* (Law on Statistics for Federal Purposes; abbreviated to BStatG) states:

> Statistics for federal purposes (federal statistics) have the task, within the federally structured overall system of official statistics, of continuously collecting, compiling, processing, presenting, and analyzing data on mass phenomena.[1]

In this context, basic principles for the production of statistics (also referred to as statistical production) apply by law or on the basis of voluntary international commitments: § 1 BStatG, for example, requires neutrality, objectivity, and professional independence. Further principles can be found in the "Quality Manual of the Statis-

---

[1]This is an unauthorized translation of "Die Statistik für Bundeszwecke (Bundesstatistik) hat im föderativ gegliederten Gesamtsystem der amtlichen Statistik die Aufgabe, laufend Daten über Massenerscheinungen zu erheben, zu sammeln, aufzubereiten, darzustellen und zu analysieren."

F. Dumpert (✉) · E. Schmidt
Federal Statistical Office of Germany, Wiesbaden, Germany
e-mail: florian.dumpert@destatis.de

tical Offices of the Federation and the Länder",[2] which in turn takes up principles from the European Statistics Code of Practice.[3] Principles 7 and 8 from the latter call for a sound methodology and appropriate statistical procedures. In particular, Principle 7.7 explicitly mentions cooperation with the scientific community:

> Statistical authorities maintain and develop cooperation with the scientific community to improve methodology, the effectiveness of the methods implemented and to promote better tools when feasible.

These and other principles are the pledge of and ensure trust in official statistics. This trust of citizens and enterprises is of essential importance. On the one hand, it facilitates or even makes possible truthful statements to the Federal Statistical Office and the Statistical Offices of the Länder, which would be ordinary state authorities if the above-mentioned principles and precepts did not apply. On the other hand, the many quality assurance steps in the statistical production process ensure that great trust is also placed in the published data.

This trust is therefore a valuable asset, and it is not for nothing that the official goal of the Federal Statistical Office has been "We uphold the trustworthiness and enhance the usefulness of our results" for several years now. Confidence in the products and high quality go hand in hand. Official statistical products are undoubtedly useful, at least as long as they are available close to the time of the survey. The larger the distance between publication and survey, the less useful such elaborately produced figures are. Official statistics must therefore be interested in the rapid production of statistics. Here, a conflict of goals between high accuracy and rapid publication appears. The conflict of goals is not new, of course, and it is common practice in national accounting, for example, to revise quickly published results at fixed points in time (on the basis of additional or better-checked information).

Statistical offices see a further starting point for enabling faster statistical production by increasing automation of statistical production. This statement can easily be misunderstood: The goal is not to eliminate the "human in the loop." Rather, the goal is to have the computer—for the case at hand, based on (data-driven) models—perform steps in the production process that are currently performed either by no one or at least not to the extent required. Many of these steps are part of Phase 5[4] of the Generic Statistical Business Process Model. To give an example, we have the following.

Example: Plausibility

When data are received in a statistical office, they are first checked for plausibility, i.e., whether the reported values are within expected parameters (ages, for example,

---

[2] https://www.destatis.de/DE/Methoden/Qualitaet/qualitaetshandbuch.html.

[3] https://ec.europa.eu/eurostat/web/products-catalogues/-/KS-02-18-142.

[4] This phase contains the subsections integrate data, classify, and code, review, and validate, edit, and impute, derive new variables and units, calculate weights, calculate aggregates, and finalise data files. Details can be found on https://statswiki.unece.org/display/GSBPM/Generic+Statistical+Business+Process+Model.

must be non-negative). If this is not the case, it is the task of a clerk to check what the true entry must be. This can be done by asking the declarant or by researching registers or the internet. In many cases, however, it is currently not possible—if only for reasons of time—to carry out these searches in such a way that a data set ultimately consists only of true values. (Whether this is even necessary for statistical production is the subject of current research and is therefore not taken up in this chapter.) In this case, it would be helpful if clerks could restrict themselves to carefully checking and plausibilizing the essential cases (for example, those with particularly large turnovers in the case of business statistics); the remaining cases would then be automatically plausibilized by the computer.

Further possible applications for automated estimations are in data preparation, e.g., by automated signatures, i.e., classifications in the technical sense.

Example: Data Preparation

An example of this is the assignment of textual data to Nomenclature statistique des Activités économiques dans la Communauté Européenne (NACE) codes, i.e. the assignment of a verbal description of an economic activity to a code classifying economic activities.

The use of models for so-called nowcasts is also conceivable. The examples mentioned could also occur in a similar way in industrial or service companies; the aspect of automation also appears relevant there. Official statistics and the economy hardly differ on this point. However, while in the economy, according to common doctrine, the market regulates whether a company will survive, official statistics exist by law and, precisely because there are no regulating market forces, are subject to their special quality requirements expressed by the principles and precepts mentioned above.

One outgrowth of the principles lies in the freedom of choice of methods and in the fact that the offices are obliged to work in a methodologically and procedurally state-of-the-art manner (Quality Manual, p. 19):

> The statistical processes for the collection, processing, and dissemination of statistics (Principles 7–10) should fully comply with international standards and guidelines and at the same time correspond to the current state of scientific research. This applies to both the methodology used and the statistical procedures applied.

Thus, the above-mentioned improvement of statistical production may (and possibly even must) be based on the use of models, or more precisely on the use of statistical (machine learning) procedures. Their quality, in turn, must be ensured and—if possible—measured against the quality of human work in the respective task fields.

7.2   Machine Learning in Official Statistics

Nowadays, a growing number of modern ML techniques are commonly used to cluster, classify, predict, and even generate tabular data, textual data, or image data. The Federal Statistical Office of Germany collects and processes various kinds of data and uses ML techniques for several tasks. Many of the tasks involved in the processing of statistics can be abstractly described as classification or regression problems, i.e., as problems from the area of supervised learning. It is therefore obvious to test machine learning methods in this field and—if the evaluation is positive—to use them in the production of statistics. The Federal Statistical Office has already made first experiences with this in the past years (see, for example, Feuerhake and Dumpert 2016, Dumpert et al. 2016, Dumpert and Beck 2017, Beck et al. 2020, Schmidt 2020, Feuerhake et al. 2020, and Dumpert and Beck 2021). At the moment, the main focus is on analyzing tabular data and textual data with ML techniques. Different ML approaches work best for different kinds of data and even in the case of data that seem to have a similar structure; not always the same ML approach as before may be superior for the specific task.

However, the use of statistical ML techniques raises another difficulty in addition to the question of which technique is best suited for a specific task:

- How to deal with the sometimes larger, sometimes smaller amount of implicit or explicit hyperparameters, which sometimes have a large, sometimes only a small influence on the performance of a method?
- Which hyperparameters should be included in the tuning that is common in this case and which should not?
- And how should tuning ideally be performed?

The quality standards of official statistics require thinking about this, also because official statistics might be obliged to be able to explain, for example, how a classification was carried out. Furthermore, results must be reproducible to a certain extent. Hence, a common optimization problem every data scientist has to face is to find the ML approach which fits the data generating process best. Usually, several approaches are tested for each data set. Common ML techniques that are used in the German Federal Statistical Office include Naïve Bayes, Elastic Net (EN), Support Vector Machine (SVM), and Decision Tree (DT) methods like Random Forest (RF) and Extreme Gradient Boosting (XGBoost). Each of these ML techniques allow for adjustment of the respective method to the specific data set with several hyperparameters. Finding the optimal hyperparameter set for a given ML method, the hyperparameter tuning, increases the search space for the best model even further.

7.3   Challenges in Tuning for the Federal Statistical Office

What to tune and how to tune it correctly? Every data scientist is faced with this question. And although literature and community propagate all kinds of standard approaches and implement them in R and PYTHON packages, there always remains the uncertainty of not having done it well, not efficiently, or not "advanced" enough. Some examples:

- Not taking the required calculation time and memory into account, the safest way to find the optimal model for the data would be to calculate ML models for all possible (combinations of) parameter settings of all available ML techniques, test all (theoretically even uncountably infinite) models, and select the best model (i.e., a "very closed meshed" grid search). Depending on the ML method and data, this is too time-consuming in most of the cases. XGBoost for example provides numerous parameters to adjust the model to the data. Tuning eight parameters of an XGBoost model with a grid search testing only six different values for each parameter results in around 1.7 million calculations (i.e., models to learn and to evaluate), which is impractical.[5] See also the discussion of model-free search methods in Sect. 4.3.

- Furthermore, even with enough time and memory, hyperparameter tuning is not straightforward and often requires expert knowledge. One reason is that there are very different types of hyperparameters. Some only refer to specific kinds of data or can only be used depending on the values of other hyperparameters. Some hyperparameters take values from a fixed set (e.g., "sampling with or without replacement" for RF[6]), for others there is a fixed range of values (e.g., "percentage of sampled data" for RF), and for others there are no restrictions at all. In the latter case, the range of values as well as the step size between the tested values have to be chosen by the user. The question arises: What is the sensitive range of values in which the impact of a hyperparameter on the model is high? In the case of other hyperparameters like the number of trees of RF, there is no optimal hyperparameter value in the sense of model quality. The model improves if the hyperparameter value is increased (or decreased) until some kind of saturation is reached (Fig. 7.1) where the model only slightly improves or does not improve anymore. Usually, in these cases, there is a trade-off between model quality and computation time. Therefore, an optimal hyperparameter value would mean that it is sufficiently high (or low) that the model cannot be improved significantly without wasting computation time. This issue is also discussed in Sect. 4.2.

- Considering the computational effort the question arises: how much does a model actually improve by hyperparameter tuning? Which ML techniques have to be tuned to lead to reasonable results? In our experience, SVMs for example seem to be very sensitive to their hyperparameters (Fig. 7.2), whereas XGBoost or RF models for example seem to provide more or less satisfactory results when default values of hyperparameters are used (Fig. 7.3[7]). In the case of a standard application,

---

[5] Assume that a model is trained and evaluated within only one second (which is not realistic for large data sets), then finding the best hyperparameter constellation out of the around 1.7 million options would take around 19 days.

[6] The authors are well aware that if theoretical results like those of Athey et al. (2019) are to be used, the type of sampling must be chosen accordingly, regardless of the tuning results.

[7] This and all the other figures in this chapter show the results for a RF or a Support Vector Machine for a function that depends on ten features.

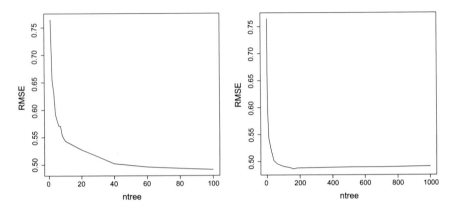

Fig. 7.1 Hyperparameter with saturation behavior

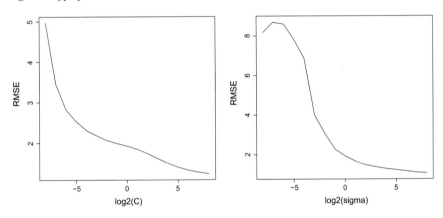

Fig. 7.2 Hyperparameters where it seems to be important to tune them (no saturation)

it might be an unnecessary overhead to spend days or weeks of computation time to improve the accuracy of a RF model in the second or third decimal place compared to the results with default values (e.g., a RMSE score without tuning (default): 0.487; tuned: 0.487; worst case: 0.494). This topic is also discussed in Chaps. 8–10, where the hyperparameter tuning processes are analyzed in the sections entitled "Analyzing the Tuning Process".

- There are several strategies to reduce the computational costs of hyperparameter tuning. One possibility is to perform a coarse-grained grid search and do a refined search in the best region of the hyperparameter space. An alternative are different search strategies that save time by only testing specific hyperparameter combinations in the high-dimensional hyperparameter space. Another possibility is to reduce the dimension of the search space. This can be done by only tuning the most promising hyperparameters or by tuning hyperparameters sequentially starting with the most promising ones and tuning only one or two hyperparameters at

Fig. 7.3 A situation where it seems to be more or less irrelevant if the hyperparameter is tuned or not

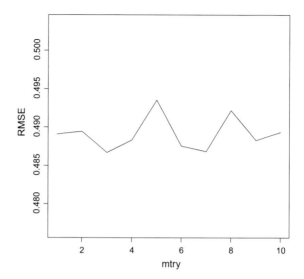

a time. State-of-the-art hyperparameter tuners such as Sequential Parameter Optimization Toolbox (SPOT) implement the concept of "active hyperparameters", i.e., the set of tuned hyperparameters can be modified. Following the latter strategies requires knowledge about the sensitivity and the interactions of hyperparameters. This book presents further approaches.

- Even hyperparameters for which a method is not very sensitive can be relevant for tuning under certain circumstances, namely if for other reasons, e.g., disclosure avoidance, a certain value may not be undercut there. In the case of tree-based methods, this applies, for example, to the minimum number of data points in the leaves. This value is analyzed in this book as the DT hyperparameter `minbucket`.
- Unfortunately, there is only limited information available (in the literature) about how hyperparameters should be tuned. ML algorithm developers usually just provide short descriptions about what a hyperparameter does. Additional information can be retrieved from a few scientific papers and online tutorials. Larger studies that address the following questions do not exist to a larger extent:
  - How much can a model be improved by tuning?
  - Which tuning strategy works best? What is the impact of tuning a specific hyperparameter of the model?
  - Does this impact vary among data sets?

The investigations Probst et al. (2019b), Probst and Boulesteix (2018), and Bischl et al. (2021b) stand out here. For own comparative investigations in the context of concrete applications, see, for example, Schmidt (2020).

Finding an optimal combination of values of the hyperparameters in a suitable sense is one goal. To show that other combinations are worse and how the different hyperparameters depend on each other is another, which may require a completely different approach. To our knowledge, the dependence of the various hyperparameters is currently limited to visual analyses, such as those presented in this book. Analytical or further empirical work on this question is, in our view, still pending.

7.4   Dealing with the Challenges

Radermacher (2020) devotes a separate section (Sect. 2.3) to quality requirements in official statistics and writes:

> 200 years of experience and a good reputation are assets as important as the profession's stock of methods, international standards and well-established routines and partnerships.

Indeed, official statistics can only provide added value because they are trusted to work in a methodologically sound manner. The same is true in the relationship between the Federal Statistical Office and its machine learning section. For this reason, the unit responsible for machine learning in the Federal Statistical Office also endeavors to carry out the hyperparameter tuning of the methods used to the best of its knowledge and—if necessary—to create transparency about it. In addition, the above-mentioned economic considerations are always involved: How long should the tuning be continued in order to perhaps still achieve a significant improvement of the models? The approaches presented in this book and further tools (cf. Bischl et al. 2021b) will support statisticians and data scientists to investigate these questions in the future, although further research is needed: The estimation of the above question (i.e., how long to tune), possibly even a statement how far away from the true optimum one is at all (at least in probability, for example, in the form of an oracle inequality), cannot be answered satisfactorily at present. Also, the effects and interactions of certain hyperparameters—see Gijsbers et al. (2021) and Moosbauer et al. (2021) for some recent considerations—have not yet been sufficiently investigated, e.g., those of the hyperparameter `respect.unordered.factors` in RF R package `ranger`. However, investigations into such issues are needed to better understand hyperparameters in the future. This can improve the basis for responsible and trustworthy use of machine learning, not only but also in official statistics.

Chapter 8
Case Study I: Tuning Random Forest (Ranger)

Thomas Bartz-Beielstein, Sowmya Chandrasekaran, Frederik Rehbach, and Martin Zaefferer

Abstract This case study gives a hands-on description of Hyperparameter Tuning (HPT) methods discussed in this book. The Random Forest (RF) method and its implementation `ranger` was chosen because it is the method of the first choice in many Machine Learning (ML) tasks. RF is easy to implement and robust. It can handle continuous as well as discrete input variables. This and the following two case studies follow the same HPT pipeline: after the data set is provided and pre-processed, the experimental design is set up. Next, the HPT experiments are performed. The R package `SPOT` is used as a "datascope" to analyze the results from the HPT runs from several perspectives: in addition to Classification and Regression Trees (CART), the analysis combines results from surface, sensitivity and parallel plots with a classical regression analysis. Severity is used to discuss the practical relevance of the results from an error-statistical point-of-view. The well proven R package `mlr` is used as a uniform interface from the methods of the packages `SPOT` and `SPOTMisc` to the ML methods. The corresponding source code is explained in a comprehensible manner.

Supplementary Information The online version contains supplementary material available at https://doi.org/10.1007/978-981-19-5170-1_8.

T. Bartz-Beielstein (✉) · S. Chandrasekaran · F. Rehbach
Institute for Data Science, Engineering and Analytics, TH Köln, Cologne, Germany
e-mail: thomas.bartzbeielstein@th-koeln.de

S. Chandrasekaran
e-mail: sowmya.chandrasekaran@th-koeln.de

F. Rehbach
e-mail: frederik.rehbach@th-koeln.de

M. Zaefferer
Bartz & Bartz GmbH and with Institute for Data Science, Engineering, and Analytics, TH Köln, Gummersbach, Germany

Duale Hochschule Baden-Württemberg Ravensburg, Ravensburg, Germany
e-mail: zaefferer@dhbw-ravensburg.de

© The Author(s) 2023
E. Bartz et al. (eds.), *Hyperparameter Tuning for Machine and Deep Learning with R*,
https://doi.org/10.1007/978-981-19-5170-1_8

187

8.1 Introduction

In this case study, the hyperparameters of the RF algorithm are tuned for a classification task. The implementation from the R package `ranger` will be used. The data set used is the Census-Income (KDD) Data Set (CID).[1]

The R package `SPOTMisc` provides a unifying interface for starting the hyperparameter-tuning runs performed in this book. The R package `mlr` is used as a uniform interface to the machine learning models. All additional code is provided together with this book. Examples for creating visualizations of the tuning results are also presented.

The hyperparameter tuning can be started as follows:

```
startCensusRun(model = "ranger")
```

This case study deals with RF. However, any ML method from the set of available methods that were discussed in Chap. 3, i.e., `glmnet`, `kknn`, `ranger`, `rpart`, `svm`, and `xgb`, can be chosen. `xgb` will be analyzed in Chap. 9.

The function `startCensusRun` performs the following steps from Table 8.1:

Table 8.1 Machine-learning hyperparameter-tuning pipeline

Step	Description, Function	Result	Details
1	getDataCensus: Data acquisition	dfCensus	Downloading the data. Compilation of a R data frame
2.1	getMlConfig: ML model and task configuration		
2.1.1	getMlrTask: Get ML Task	task	ml task
2.1.2	getModelConf: Model configuration	cfg	Model
2.1.3	getMlrResample: Split Data into Training and Test Data	data	Partitioned data
2.2	getObjf: Objective function	objf	Objective function
3	spot: Hyperparameter tuning	result	Result list
4	evalParamCensus: Evaluate on test data	Score	Metrics

---

[1] The data from CID is historical. It includes wording or categories regarding people which do not represent or reflect any views of the authors and editors.

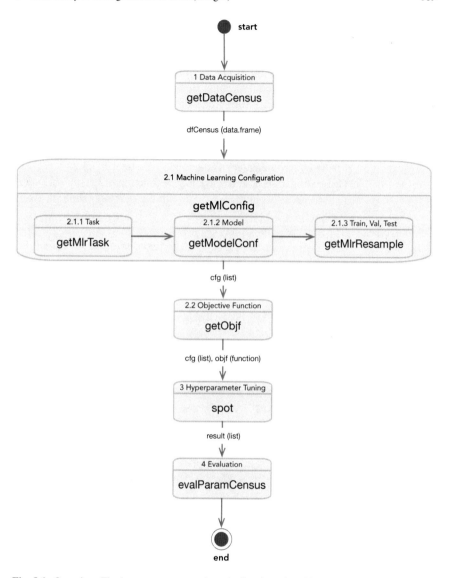

Fig. 8.1 Overview. The hyperparameter-tuning pipeline introduced in this chapter comprehends four main steps. After the data acquisition (getDataCensus), the ML model is configured (getMlConfig) and the objective function (getObjf) is specified. The hyperparameter tuner SPOT is called (spot) and finally, results are evaluated (evalParamCensus). The ML configuration via getMlConfig combines the results from three subroutines, i.e., getMlrTask, getModelConf, and getMlrResample

1 Data:	Acquisition and preparation of the CID data set. The function getDataCensus is called to perform these steps; see Sect. 8.2.1.
2.1 Design:	The experimental design is set up. This step includes the specification of measures, the configuration of the hyperparameter tuner, and the configuration of the ML model. Calling the function getMlConfig executes the subroutines 2.1.1 until 2.1.3:
2.1.1 Task:	Definition of a ML task. The function getMlrTask performs this step. It results in an mlr task object; see Sect. 8.3.2.1.
2.1.2 Config:	Hyperparameter configuration. The function getModelConf sets up the hyperparameters of the model; see Sect. 8.3.2.2.
2.1.3 Split:	Generating training and test data. The function getMlrResample is used here. It returns a list with the corresponding data sets; see Sect. 8.3.2.3.
2.2 Objective:	The objective function is defined via getObjf; see Sect. 8.4.4.
3 Tuning:	The hyperparameter tuner, i.e., the function spot, is called. See Sect. 8.5.
4 Evaluation:	Evaluation on test data. To evaluate the results, the function evalParamCensus can be used; see Sect. 8.6.2. These steps are illustrated in Fig. 8.1.

8.2 Data Description

8.2.1 The Census Data Set

For the investigation, we choose the CID, which is made available, for example, via the UCI ML Repository.[2] For our investigation, we will access the version of the data set that is available via the platform openml.org under the data record ID 45353 (Vanschoren et al. 2013). This data set is an excerpt from the current population surveys of 1994 and 1995, compiled by the U.S. Census Bureau. It contains $n = 299,285$ observations with 41 features on demography and employment. The data set is comparatively large, has many categorical features with many levels, and fits well with the field of application of official statistics.

The CID data set suits our research questions well since it is comparatively large and has many categorical features. Several of the categorical features have a broad variety of levels. The data set can be easily used to generate different classification and regression problems.

The data preprocessing consists of the following steps:

- Feature 24 (instance weight MARSUPWT) is removed. This feature describes the number of persons in the population who are represented by the respective obser-

---

[2] https://archive.ics.uci.edu/ml/datasets/Census-Income+(KDD).

vation. This is relevant for data understanding, but should not be an input to the ML models.

- Several features are encoded as numerical (integer) variables, but are in fact categorical. For example, feature 3 (industry code ADTIND) is encoded as an integer. Since the respective integers represent discrete codes for different sectors of industry, they have no inherent order and should be encoded as categorical features.
- The data set sometimes contains NA values (missing data). These NA values are replaced before modeling. For categorical features, the most frequently observed category is imputed (mode). For integer features, the median is imputed, and for real-valued features the mean.
- As the only model investigated in this book, xgboost is not able to work directly with categorical features. This becomes relevant for the experiments in Chap. 12. In that case (only for xgboost), the categorical data features are transferred into a dummy coding. For each category of the categorical feature, a new binary feature is created, which specifies whether an observation is of the respective category or not.
- Finally, we split the data randomly into test data (40% of the observations) and training data (60%).

In addition to these general preprocessing steps, we change the properties of the data set for individual experiments to cover our various hypotheses (esp. in Chap. 12). Arguably, we could have done this by using completely different data sets where each set covers different objects of investigation (i.e., different numbers of features or different m). We decided to stick to a single data set and vary it instead of generating new, comparable data sets with different properties. This allows us to reasonably compare results between the individual variations. This way, we generate multiple data sets that cover different aspects and problems in detail. While they all derive from the same data set (CID), they all have different characteristics: Number of categorical features, number of numerical features, cardinality, number of observations, and target variable. These characteristics can be quantified with respect to difficulty as discussed in Sect. 12.5.4.

In detail, we vary:

Target: The original target variable of the data set is the income class (below/above 50 000 USD). We choose *age* as the target variable instead. For classification experiments, age will be discretized, into two classes: age < 40 and age >= 40. For regression, age remains unchanged. This choice intends to establish comparability between both experiment groups (classification, regression).

cardinality: The number of categories (cardinality). To create variants of the data set with different cardinality of categorical features, we merge categories into new, larger categories. For instance, for feature 35 (country of birth self PENATVTY) the country of origin is first merged by combining all countries from a specific continent. This reduces the cardinality, with 6 remaining cate-

	gories (medium cardinality). For a further reduction (low cardinality) to three categories, the data is merged into the categories unknown, US, and abroad. Similar changes to other features are documented in the source code. For our experiments, this pre-processing step results in data sets with the levels of cardinality: low (up to 15 categories), medium (up to 24 categories), and high (up to 52 categories).
nnumericals:	Number of numerical features (nnumericals). To change the number of features, individual features are included or removed from the data set. This is done separately for categorical and numerical features and results in four levels for nnumericals (low: 0, medium: 4, high: 6, complete: 7).
nfactors:	Number of categorical features (nfactors). Correspondingly, we receive four levels for nfactors (low: 0, medium: 8, high: 16, complete: 33). Note, that these numbers become somewhat reduced, if cardinality is low (low: 0, medium: 7, high: 13, complete: 27). The reason is that some features might become redundant when merging categories.
n:	Number of observations (n). To vary n, observations are randomly sampled from the data set. We test five levels on a logarithmic scale from 10^4 to 10^5: 10 000, 17 783, 31 623, 56 234, and 100 000. In addition, we conduct a separate test with the complete data set, i.e., 299 285 observations.

To keep results comparable, most case studies in this book (Chaps. 9, 10, and 12) use the same data preprocessing of the CID data set. Only Chap. 12 considers several variations of the CID data set simultaneously.

Background: Implementation Details

The function getDataCensus from the package SPOTMisc uses the functions setOMLConfig and getOMLDataSet from the R package OpenML, i.e., the CID can also be downloaded as follows:

```
OpenML::setOMLConfig(cachedir = "oml.cache")
dataOML <- OpenML::getOMLDataSet(4535)$data
```

While not strictly necessary, it is a good idea to set a permanent cache directory for Open Machine Learning (OpenML) data set. Otherwise, every new experiment will redownload the data set, taxing the OpenML servers unnecessarily.

Information about the 42 columns of the CID data set is shown in Table 8.2.

Table 8.2 CID data set

Var	Type, factor levels	Example data
V1:	num:	73 58 18 9 10 48 42 28 47 34 ...
V2:	Factor w/ 9 levels "Federal government", ...:	4 7 4 4 4 5 5 5 2 5 ...
V3:	num:	0 4 0 0 0 40 34 4 4 43 4 ...
V4:	num:	0 34 0 0 0 10 3 40 26 37 ...
V5:	Factor w/ 17 levels "10th grade", ...:	13 17 1 11 11 17 10 13 17 17 ...
V6:	num:	0 0 0 0 0 1200 0 0 876 0 ...
V7:	Factor w/ 3 levels "College or university", ...:	3 3 2 3 3 3 3 3 3 3 ...
V8:	Factor w/ 7 levels "Divorced", "Married-A F spouse present", ...:	7 1 5 5 5 3 3 5 3 3 ...
V9:	Factor w/ 24 levels "Agriculture", ...:	15 5 15 15 15 7 8 5 6 5 ...
V10:	Factor w/ 15 levels "Adm support including clerical", ...:	7 9 7 7 7 11 3 5 1 6 ...
V11:	Factor w/ 5 levels "Amer Indian Aleut or Eskimo", ...:	5 5 2 5 5 1 5 5 5 5 ...
V12:	Factor w/ 10 levels "All other", "Central or South American", ...:	1 1 1 1 1 1 1 1 1 1 ...
V13:	Factor w/ 2 levels "Female", "Male":	1 2 1 1 1 1 2 1 1 2 ...
V14:	Factor w/ 3 levels "No", "Not in universe", ...:	2 2 2 2 2 1 2 2 1 2 ...
V15:	Factor w/ 6 levels "Job leaver", ...:	4 4 4 4 4 4 4 2 4 4 ...
V16:	Factor w/ 8 levels "Children or Armed Forces", ...:	3 1 3 1 1 2 1 7 2 1 ...
V17:	num :	0 0 0 0 0 ...
V18:	num :	0 0 0 0 0 0 0 0 0 0 ...
V19:	num :	0 0 0 0 0 0 0 0 0 0 ...
V20:	Factor w/ 6 levels "Head of household", ...:	5 1 5 5 5 3 3 6 3 3 ...
V21:	Factor w/ 6 levels "Abroad", "Midwest", ...:	4 5 4 4 4 4 4 4 4 4 ...
V22:	Factor w/ 51 levels "?", "Abroad", ...:	37 6 37 37 37 37 37 37 37 37 ...
V23:	Factor w/ 38 levels "Child <18 ever marr not in subfamily", ...:	30 21 8 3 3 37 21 36 37 21 ...
V24:	Factor w/ 8 levels "Child 18 or older", ...:	7 5 1 3 3 8 5 6 8 5 ...
V25:	num:	1700 1054 992 1758 1069 ...
V26:	Factor w/ 10 levels "?", "Abroad to MSA", ...:	1 4 1 6 6 1 6 1 1 6 ...
V27:	Factor w/ 9 levels "?", "Abroad", ...:	1 9 1 7 7 1 7 1 1 7 ...
V28:	Factor w/ 10 levels "?", "Abroad", ...:	1 10 1 8 8 1 8 1 1 8 ...
V29:	Factor w/ 3 levels "No", "Not in universe under 1 year old", ...:	2 1 2 3 3 2 3 2 2 3 ...
V30:	Factor w/ 4 levels "?", "No", "Not in universe", ...:	1 4 1 3 3 1 3 1 1 3 ...
V31:	num:	0 1 0 0 0 1 6 4 5 6 ...
V32:	Factor w/ 5 levels "Both parents present", ...:	5 5 5 1 1 5 5 5 5 5 ...
V33:	Factor w/ 43 levels "?", "Cambodia", ...:	41 41 42 41 41 32 41 41 41 41 ...
V34:	Factor w/ 43 levels "?", "Cambodia", ...:	41 41 42 41 41 41 41 41 41 41 ...
V35:	Factor w/ 43 levels "?", "Cambodia", ...:	41 41 42 41 41 41 41 41 41 41 ...
V36:	Factor w/ 5 levels "Foreign born- Not a citizen of U S", ...:	5 5 1 5 5 5 5 5 5 5 ...
V37:	num:	0 0 0 0 0 2 0 0 0 0 ...
V38:	Factor w/ 3 levels "No", "Not in universe", ...:	2 2 2 2 2 2 2 2 2 2 ...
V39:	num:	2 2 2 0 0 2 2 2 2 2 ...
V40:	num:	0 52 0 0 0 52 52 30 52 52 ...
V41:	num:	95 94 95 94 94 95 94 95 95 94 ...
V42:	Factor w/ 2 levels "-50000.", "50000+.":	1 1 1 1 1 1 1 1 1 1 ...

8.2.2 *getDataCensus: Getting the Data from OML*

The CID data set can be configured with respect to the target variable, the task, and the complexity of the data (e.g., number of samples, cardinality). The following variables are defined:

```
target <- "age"
task.type <- "classif"
nobs <- 1e4
nfactors <- "high"
nnumericals <- "high"
cardinality <- "high"
data.seed <- 1
cachedir <- "oml.cache"
```

These variables will be passed to the function getDataCensus to obtain the data frame dfCensus (Fig. 8.2). The function getDataCensus is used to get the OML data (from cache or from server). The arguments target, task.type, nobs, nfactors, nnumericals, cardinality and cachedir can be used, see Table 8.3.

```
dfCensus <- getDataCensus(
  task.type = task.type,
  nobs = nobs,
  nfactors = nfactors,
  nnumericals = nnumericals,
  cardinality = cardinality,
  data.seed = data.seed,
  cachedir = cachedir,
  target = target
)
```

The dfCensus data set used in the case studies has 10 000 observations of 23 variables, which are shown in Table 8.4.

Table 8.3 Parameters used to get the CID data set. A detailed description can be found in Sect. 8.2.1

Parameter	Value used in the case studies	Description
Target	"age"	Target variable. Age smaller or larger 40 years
Cachedir	"oml.cache"	Location of the cached data
task.type	"classif"	Classification task. The target is used for defining the classes
Nobs	1e4	The complete data set has 299, 285 observations. nobs observations are randomly sampled
nfactors	"high"	Number of categorical features
nnumericals	"high"	Number of numerical features
Cardinality	"high"	Number of categories
data.seed	1	Seed used for sampling nobs observations

Table 8.4 The dfCensus data set

Parameter	Type	Storage levels	Mode, Example	Description
Capital_gains	Num	Double	0	min: 0. max: 9.9999×10^4. 3.56 % have capital gains
Capital_losses	Num	Double	0	min: 0. max: 3900. 1.66 % have capital losses
Sivdends_from_stocks	Num	Double	0	min: 0. max: 9.9999×10^4. 9.96 % have dividends from stock
Num_persons_-worked_for_employer	Num	Integer	0	min: 0. max: 6.
Wage_per_hour	Num	Double	0	min: 0. max: 6800.
Weeks_worked_in_year	Num	Integer	0	min: 0. max: 52.
Class_of_worker	Factor	9	"Federal government"	
Industry_code	Factor	51	"0"	
Occupation_code	Factor	47	"0"	
Education	Factor	17	"10th grade"	
Marital_status	Factor	7	"Divorced"	
Major_industry_code	Factor	24	"Agriculture"	
Major_occupation_code	Factor	15	"Adm support including clerical"	
Race	Factor	5	"Amer Indian Aleut or Eskimo"	
Hispanic_origin	Factor	10	"All other"	
Sex	factor	2	"Female", "Male"	
Tax_filer_status	Factor	6	"Head of household"	
Detailed_household_and-_family_stat	Factor	29	"Child < 18 ever marr not in subfamily"	
Detailed_household_summary_in_household	Factor	8	"Child 18 or older"	
Country_of_birth_self	factor	42	"?", "Cambodia"	
Citizenship	Factor	5	"Foreign born- Not a citizen of U S"	
Income_class	Factor	2	"-50000.", "50000+."	
Target	Factor	2	"FALSE", "TRUE"	

> **!** **Attention: Outliers and Inconsistent Data**

- Target: The values of the target variable are not equally balanced, because 61.27% of the values are TRUE, i.e., older than 40 years.
- The numerical variables num_persons_worked_for_employer and weeks_worked_in_year can be treated as an integers.
- The factor income_class can be treated as a logical value.
- Summaries of the numerical variables:

```
summary(dfCensus[,sapply(dfCensus, is.numeric)])
```

```
##   wage_per_hour        capital_gains       capital_losses      divdends_from_stocks
##   Min.   :   0.00   Min.   :    0.0    Min.   :   0.00    Min.   :    0.0
##   1st Qu.:   0.00   1st Qu.:    0.0    1st Qu.:   0.00    1st Qu.:    0.0
##   Median :   0.00   Median :    0.0    Median :   0.00    Median :    0.0
##   Mean   :  51.46   Mean   :  452.1    Mean   :  31.22    Mean   :  202.2
##   3rd Qu.:   0.00   3rd Qu.:    0.0    3rd Qu.:   0.00    3rd Qu.:    0.0
##   Max.   :6800.00   Max.   :99999.0    Max.   :3900.00    Max.   :99999.0
##   num_persons_worked_for_employer weeks_worked_in_year
##   Min.   :0.000                    Min.   : 0.00
##   1st Qu.:0.000                    1st Qu.: 0.00
##   Median :1.000                    Median : 6.00
##   Mean   :1.922                    Mean   :22.75
##   3rd Qu.:4.000                    3rd Qu.:52.00
##   Max.   :6.000                    Max.   :52.00
```

- capital_gains and divdends_from_stocks share the same upper limit: 9.9999×10^4, which appears to be an artificial upper limit.
- Wage per Hour: There is one entry with 6800, but income class –50000.

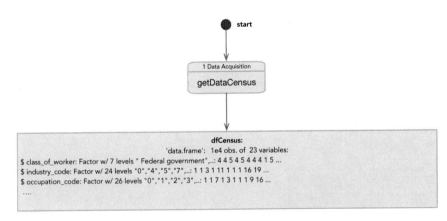

Fig. 8.2 Step 1 of the hyperparameter-tuning pipeline introduced in this chapter: the data acquisition (getDataCensus) generates the data set dfCensus, which is a subset of the full CID data set presented in Table 8.2, because the parameter setting nobs = 1e4, nfactors = "high", nnumericals = "high", and cardinality = "high" was chosen

8.3   Experimental Setup and Configuration of the Random Forest Model

8.3.1   *getMlConfig*: *Configuration of the ML Models*

Since we are considering a binary classification problem (age, i.e., young versus old), the mlr task.type is set to classif. Random forests ("ranger") will be used for classification.

```
model <- "ranger"
cfg <- getMlConfig(
  target = target,
  model = model,
  data = dfCensus,
  task.type = task.type,
  nobs = nobs,
  nfactors = nfactors,
  nnumericals = nnumericals,
  cardinality = cardinality,
  data.seed = data.seed,
  prop = 2 / 3
)
```

As a result from calling getMlConfig, the list cfg is available. This list has 13 entries, that are summarized in Table 8.5.

Table 8.5 Configuration list with 13 entries as a result from calling getMlConfig

Parameter	Value	Description
Learner	"classif.ranger"	Learner
Tunepars	"num.trees", "mtry", "sample.fraction", "replace", and "respect.unordered.factors"	The hyperparameters of the model
Defaults		Default hyperparameter settings of the tunepars
Lower		Lower bounds of the hyperparameters
Upper		Upper bounds of the hyperparameters
Type	"numeric", "integer", or "factor"	Hyperparameter variable types
Fixpars	–	Fixed hyperparameters
Factorlevels	Levels of each factor variable	
Transformations	Applied transformations	
Dummy	Dummy encoding	Used by xgboost
Relpars	–	Parameters relative to others
Task	mlr task object	
Resample	Resampling strategy from mlr	

8.3.2 Implementation Details: *getMlConfig*

The function getMlConfig combines results from the following functions

- getMlrTask
- getModelConf
- getMlrResample

The functions will be explained in the following (Fig. 8.3).

8.3.2.1 getMlrTask: Problem Design and Definition of the Machine Learning Task

The target variable of the data set is age (age below or above 40 years). The problem design describes target and task type, the number of observations, as well as the number of factorial, numerical, and cardinal variables. The data seed can also be specified here.

```
task <- getMlrTask(dataset = dfCensus,
task.type = "classif",
```

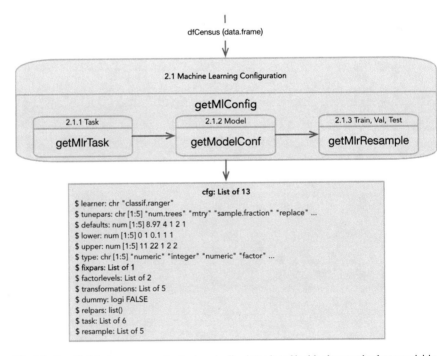

Fig. 8.3 Step 2 of the hyperparameter-tuning pipeline introduced in this chapter: the data acquisition (getMlConfig) generates the list cfg

```
data.seed = 1)
```

The function `getMlrTask` is an interface to the function `makeClassifTask` from the `mlr` package. The resulting `task` "encapsulates the data and specifies—through its subclasses—the type of the task. It also contains a description object detailing further aspects of the data." (Bischl et al. 2016).

Background: `getMlrTask` Implementation

The data set `dfCensus` is passed to the function `getMlrTask`, which computes an `mlr` `task` as shown below:

```
getMlrTask <- function(dataset,
                       task.type = "classif",
                       data.seed = 1) {
  target <- "target"
  task.nobservations <- nrow(dataset)
  task.nfeatures <- ncol(dataset)
  task.numericals <- lapply(dataset, class) != "factor"
  task.numericals[target] <- FALSE
  task.factors <- lapply(dataset, class) == "factor"
  task.factors[target] <- FALSE
  task.nnumericals <- sum(task.numericals)
  task.nfactors <- sum(task.factors)
  task.nlevels <-
    as.numeric(lapply(dataset, function(x) {
      length(unique(x))
    }))
  task.nlevels[!task.factors] <- 0
  task.nlevels.max <- max(task.nlevels)
  task <- makeClassifTask(data = dataset, target = target)
  task <- impute(task,
    classes = list(
      factor = imputeMode(),
      integer = imputeMedian(),
      numeric = imputeMean()
    )
  )$task
  return(task)
}
```

Because the function `getMlrTask` generates an `mlr` Task instance, its elements can be accessed with `mlr` methods, i.e., functions from `mlr` can be applied to the Task `task`. For example, the feature names that are based on the data can be obtained with the `mlr` function `getTaskFeatureNames` as follows:

```
head(getTaskFeatureNames(task))

## [1] "class_of_worker" "industry_code"   "occupation_code" "education"
## [5] "wage_per_hour"   "marital_status"
```

The Task task provides the basis for the the information that is needed to perform the hyperparameter tuning. Additional information is generated by the functions getModelConf, that is presented next.

8.3.2.2 getModelConf: Algorithm Design—Hyperparameters of the Models

The function getModelConf generates a list with the entries learner, tunepars, defaults, lower, upper, type, fixpars, factorlevels, transformations, dummy, relpars, task, and resample that are summarized in Table 8.5.

The ML configuration list modelCfg contains information about the hyperparameters of the ranger model; see Table 8.6.

```
nFeatures <- sum(task$task.desc$n.feat)
modelCfg <- getModelConf(
  task.type = task.type,
  model = model,
  nFeatures = nFeatures
)
```

Background: Model Information from getModelConf

The information about the ranger hyperparameters, their ranges and types, is compiled as a list. It is accessible via the function getModelConf. This function manages the information about the ranger model as follows:

Table 8.6 Ranger hyperparameter. N_{Feats} denotes the output from getTaskNFeats(task)

Variable	Name	Type	Default	Upper	Lower	Trans
x_1	num.trees	Numeric	8.965784	0	11	2pow_round
x_2	mtry	Integer	4	1	22	id
x_3	sample.fraction	Numeric	1	0.1	1	id
x_4	Replace	Factor	2	1	2	id
x_5	respect.unordered.factors	Factor	1	1	2	id

```
learner <- paste(task.type, "ranger", sep = ".")
tunepars <- c(
  "num.trees",
  "mtry",
  "sample.fraction",
  "replace",
  "respect.unordered.factors"
)
defaults <- c(
  log(500, 2), floor(sqrt(nFeatures)), 1,
  2, 1
)
lower <- c(0, 1, 0.1, 1, 1)
upper <- c(11, nFeatures, 1, 2, 2)
type <- c(
  "numeric", "integer", "numeric", "factor",
  "factor"
)
fixpars <- list(num.threads = 1)
factorlevels <-
  list(
    respect.unordered.factors = c(
      "ignore",
      "order", "partition"
    ),
    replace = c(FALSE, TRUE)
  )
transformations <- c(
  trans_2pow_round, trans_id, trans_id,
  trans_id, trans_id
)
dummy <- FALSE
relpars <- list()
```

Similar information is provided for every ML model. Note: This list is independent from `mlr`, i.e., it does not use any `mlr` classes.

8.3.2.3 `getMlrResample`: Training and Test Data

The function `getMlrResample` is the third and last subroutine used by the function `getMlConfig`. It takes care of the partitioning of the data into training data, $X^{(train)}$, and test data, $X^{(test)}$, based on `prop`. The function `getMlrResample` from the package `SPOTMisc` is an interface to the `mlr` function `makeFixedHoldout Instance`, which generates a fixed holdout instance for resampling.

```
rsmpl <- getMlrResample(task=task,
dataset = dfCensus,
data.seed = 1,
prop = 2/3)
```

rsmpl specifies the training data set, $X^{(\text{train})}$, which contains prop $= 2/3$ of the data and the testing data set, $X^{(\text{test})}$ with the remaining $1 - \text{prop} = 1/3$ of the data. It is implemented as a list, the central components are lists of indices to select the members of the corresponding train or test data sets.

Background: getMlrResample

Information about the data split are stored in the cfg list as cfg$resample. They can also be computed directly using the function getMlrResample. This function computes an mlr resample instance generated with the function makeFixedHoldoutInstance.

```
getMlrResample <- function(task,
                           dataset,
                           data.seed = 1,
                           prop = NULL) {
  set.seed(data.seed)
  train.id <- sample(1:getTaskSize(task),
                     size = getTaskSize(task) * prop,
                     replace = FALSE)
  test.id <- (1:getTaskSize(task))[-train.id]
  rsmpl = makeFixedHoldoutInstance(train.id,
                                   test.id,
                                   getTaskSize(task))
  return(rsmpl)
```

The function getMlrResample instantiated an mlr resampling strategy object from the class makeResampleInstance. This mlr class encapsulates training and test data sets generated from the data set for multiple iterations. It essentially stores a set of integer vectors that provide the training and testing examples for each iteration. (Bischl et al. 2016). The first entry, desc, describes the split between training and test data and its properties, e.g., what to predict during resampling: "train", "test" or "both" sets. The second entry, size, stores the size of the data set to resample. The third and fourth elements are lists with the training and test indices, i.e., for 6666 indices for the $X^{(\text{train})}$ data set and 3334 indices for the $X^{(\text{test})}$ data set. These indices will be used for all iterations. The last element is optional and encodes whether specific iterations "belong together" (Bischl et al. 2016).

str(rsmpl)

```
## List of 5 ## $
desc      :List of 7 ##   ..$ split      : num 0.667 ## ..$ id : chr
"holdout" ## ..$ iters     : int 1 ## ..$ predict    : chr "test"
##    ..$ stratify   : logi FALSE ## ..$ fixed      : logi FALSE ##
..$ blocking.cv: logi FALSE ## ..- attr(*, "class")= chr [1:2]
"HoldoutDesc" "ResampleDesc" ## $ size : int 10000 ##   $
train.inds:List of 1 ##   ..$ : int [1:6666] 1017 8004 4775 9725
8462 4050 8789 1301 8522 1799 ... ## $ test.inds :List of 1 ## ..$
: int [1:3334] 1 10 11 12 13 17 20 23 25 28 ... ##   $ group : Factor
w/ 0 levels: ##   - attr(*, "class")= chr "ResampleInstance"
```

> **Important: Training and Test Data**

mlr's function `resample` requires information about the test data, because it manages the train and test data partition internally. Usually, it is considered "best practice" in ML not to pass the test set to the ML model. To the best of our knowledge, this is not possible in `mlr`.

Therefore, the full data set (training and test data) with `nobs` $= 10^4$ observations is passed to the `resample` function. Because `mlr` is an established R package, we trust the authors that `mlr` keeps training and test data separately.

An additional problem occurs if the test data set, $X^{(\text{test})}$, contains data with unknown labels, i.e., factors with unknown levels. If these unknown levels are passed to the trained model, predictions cannot be computed.

8.4   Objective Function (Model Performance)

8.4.1   Performance Measures

The evaluation of hyperparameter values requires a measure of quality, which determines how well the resulting models perform. For the classification experiments, we use Mean Mis-Classification Error (MMCE) as defined in Eq. (2.2). The hyperparameter tuner Sequential Parameter Optimization Toolbox (SPOT) uses these MMCE on the test data set to determine better hyperparameter values.

In addition to MMCE, we also record run time (overall run time of a model evaluation, run time for prediction, run time for training). To mirror a realistic use case, we specify a fixed run time budget for the tuner. This limits how long the tuner may take to find potentially optimal hyperparameter values.

For a majority of the models, the run time of a single evaluation (training + prediction) is hard to predict and may easily become excessive if parameters are chosen poorly. In extreme cases, the run time of a single evaluation may exceed drastically the planned run time. In such a case, there would be insufficient time to test different hyperparameter values. To prevent this, we specify a limit for the run time of a single evaluation, which we call `timeout`. When the `timeout` is exceeded by the model, the evaluation will be aborted. During the experiments, we set the timeout to a twentieth of the tuner's overall run time budget.

```
timebudget <- 60 ## secs
timeout <- timebudget / 20
```

Exceptions are the experiments with Decision Tree (DT) (`rpart`): Since `rpart` evaluates extremely quickly, (in our experiments: usually much less than a second) the timeout is not required. In fact, using the `timeout` would add considerable overhead to the evaluation time in this case.

The HPT task can be parallelized by specifying `nthread` values larger than one. Only one thread was used in the experiment. In addition to Root Mean Squared Error (RMSE) and MMCE, we also record run time (overall run time of a model evaluation, run time for prediction, run time for training). Several alternative metrics can be specified.

Example: Changing the loss function

For example, `logloss` can be selected as follows:

```
if (task.type == "classif") {
  fixpars <- list(
    eval_metric = "logloss",
    nthread = 1
  )
} else {
  fixpars <- list(
    eval_metric = "rmse",
    nthread = 1
  )
}
```

8.4.2 Handling Errors

If the evaluation is aborted (e.g., due to `timeout` or in case of some numerical instability), we still require a quality value to be returned to the tuner, so that the search can continue. This return value should be chosen, so that, e.g., additional evaluations with high run times are avoided. At the same time, the value should be on a similar scale as the actual quality measure, to avoid a deterioration of the underlying surrogate model. To achieve this, we return the following values when an evaluation aborts.

- Classification: Model quality for simply predicting the mode of the training data.
- Regression: Model quality for simply predicting the mean of the training data.

8.4.3   Imputation of Missing Data

The imputation of missing values can be implemented using built-in methods from `mlr`. These imputations are based on the hyperparameter types: factor variables will use `imputeMode`, integers use `imputeMedian`, and numerical values use `imputeMean`.

❗ Important: Imputation

There are two situations when imputation can be applied:

1. Missing data, i.e., CID data are incomplete. This imputation can be handled by the `mlr` methods described in this section.
2. Missing results, i.e., performance values of the ML method such as loss or accuracy. This imputation can be handled by `spot`.

8.4.4   `getObjf`: The Objective Function

After the ML configuration is compiled via `getMlConfig`, the objective function has to be generated.

```
objf <- getObjf(
  config = cfg,
  timeout = timeout
)
```

The `getObjf` compiles information from the `cfg` and information about the budget (`timeout`) (Fig. 8.4).

Background: `getObjf` as an Interface to `mlr`'s `resample` function

Note, in addition to hyperparameter information, `cfg` includes information about the `mlr task`. `getObjf` calls the `mlr` function `makeLearner`. The information is used to execute the `resample` function, which fits a model specified by `learner` on a `task`. Predictions and performance measurements are computed for all training and testing sets specified by the resampling method (Bischl et al. 2016).

A simplified version that implements the basic elements of the function `getObjf`, is shown below. After the parameter names are set, the parameter transformations are performed and the complete set of parameters is compiled: this includes converting

Fig. 8.4 Step 3 of the
hyperparameter-tuning
pipeline introduced in this
chapter: (getObjf)
generates the function objf

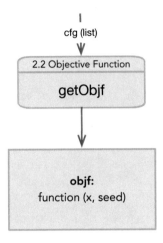

integer levels to factor levels for categorical parameters, setting fixed parameters
(which are not tuned, but are also not set to default value), and setting parameters in relation to other parameters (e.g., minbucket relative to minsplit). Next, the
learner lrn is generated via mlr's function makeLearner, and the measures are
defined. Here, the fixed set mmce, timeboth, timetrain, and timepredict
are used. After setting the Random Number Generator (RNG) seed, the mlr function
resample is called. The function resample fits a model specified by the learner
on a task and calculates performance measures for all training sets, $X^{(train)}$, and all
test sets, $X^{(test)}$, specified by the resampling instance config$resample that was
generated with the function getMlrResample as described in Sect. 8.3.2.3.

```r
getObjf <- function(config, timeout = 3600) {
  objfun <- function(x, seed) {
    params <- as.list(x)
    names(params) <- config$tunepars
    for (i in 1:length(params)) {
      params[[i]] <- config$transformations[[i]](params[[i]])
    }
    params <- int2fact(params, config$factorlevels)
    params <- c(params, config$fixpars)
    nrel <- length(config$relpars)
    for (i in 1:nrel) {
      params[names(config$relpars)[i]] <-
        with(params, eval(config$relpars[[i]]))
    }
    lrn <- makeLearner(config$learner, par.vals = params)
    measures <- list(mmce, timeboth, timetrain, timepredict)
    set.seed(seed)
    res <- resample(lrn,
      config$task,
      config$resample,
      measures = measures
```

```
    )
    timestamp <- as.numeric(Sys.time())
    return(matrix(c(res$aggr, timestamp), 1))
  }
  objvecf <- function(x, seed) {
    res <- NULL
    for (i in 1:nrow(x)) {
      res <- rbind(res, objfun(x[i, , drop = FALSE], seed[i]))
    }
    return(res)
  }
}
```

The return value, res, of the objective function generated with getObjf was evaluated on the test set, $X^{(test)}$.

names(res$aggr)

```
## [1] "mmce.test.mean"        "timeboth.test.mean"     "timetrain.test.mean"
## [4] "timepredict.test.mean"
```

No explicit validation set, $X^{(val)}$, is defined during the HPT procedure.

Importantly, randomization is handled by spot by managing the seed via spot's seedFun argument. The seed management guarantees that two different hyperparameter configurations, λ_i and λ_j, are evaluated on the same test data $X^{(test)}$. But if the same configuration is evaluated a second time, it will receive a new test data set.

8.5 spot: Experimental Setup for the Hyperparameter Tuner

The R package SPOT is used to perform the actual hyperparameter tuning (optimization). The hyperparameter tuner itself has parameters such as kind and size of the initial design, methods for handling non-numerical data (e.g., Inf, NA, NaN), the surrogate model and the optimizer, search bounds, number of repeats, methods for handling noise.

Because the generic SPOT setup was introduced in Sect. 4.5, this section highlights the modifications of the generic setup that were made for the ML runs.

The third step of the hyperparameter-tuning pipeline as shown in Fig. 8.5 starts the SPOT hyperparameter tuner.

```r
result <- spot(
  x = NULL,
  fun = objf,
  lower = cfg$lower,
  upper = cfg$upper,
  control = list(
    types = cfg$type,
    time = list(maxTime = timebudget / 60),
    noise = TRUE,
    OCBA = TRUE,
    OCBABudget = 3,
    seedFun = 123,
    designControl = list(
      replicates = Rinit,
      size = initSizeFactor * length(cfg$lower)
    ),
    replicates = 2,
    funEvals = Inf,
    modelControl = list(
      target = "ei",
      useLambda = TRUE,
      reinterpolate = TRUE
    ),
    optimizerControl = list(funEvals = 200 * length(cfg$lower)),
    multiStart = 2,
    parNames = cfg$tunepars,
    yImputation = list(
      handleNAsMethod = handleNAsMean,
      imputeCriteriaFuns = list(is.infinite, is.na, is.nan),
      penaltyImputation = 3
    )
  )
)
```

The result from the spot run is the result list, which can be written to a file. The full R code for running this case study is shown Sect. 8.10 and the SPOT parameters are listed in Table 8.7.

Background: Implementation details of the function spot

The initial design is created by Latin Hypercube Sampling (LHS) (Leary et al. 2003). The size of that design (number of sampled configurations of hyperparameters) corresponds to $2 \times k$. Here, k is the number of hyperparameters.

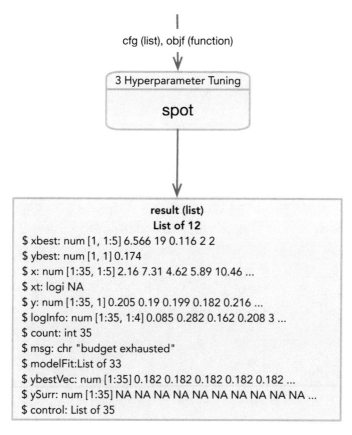

cfg (list), objf (function)

3 Hyperparameter Tuning

spot

result (list)
List of 12
$ xbest: num [1, 1:5] 6.566 19 0.116 2 2
$ ybest: num [1, 1] 0.174
$ x: num [1:35, 1:5] 2.16 7.31 4.62 5.89 10.46 ...
$ xt: logi NA
$ y: num [1:35, 1] 0.205 0.19 0.199 0.182 0.216 ...
$ logInfo: num [1:35, 1:4] 0.085 0.282 0.162 0.208 3 ...
$ count: int 35
$ msg: chr "budget exhausted"
$ modelFit:List of 33
$ ybestVec: num [1:35] 0.182 0.182 0.182 0.182 0.182 ...
$ ySurr: num [1:35] NA NA NA NA NA NA NA NA NA NA ...
$ control: List of 35

Fig. 8.5 The hyperparameter-tuning pipeline: the hyperparameter tuner SPOT is called (spot)

Table 8.7 SPOT parameters used for ML hyperparameter tuning. Parameters, that are implemented as lists are described in Table 8.8. This table shows only parameters that were modified for the ML and DL hyperparameter-tuning tasks. The full list is shown in Table 4.2

Parameter	Value	Description
x	x0	Starting point
fun	objf	Objective function as described in Sect. 8.4.4
lower	cfg$lower	Lower bound
upper	cfg$upper	Upper bound
control	list	See description in Table 8.8

Table 8.8 SPOT control list parameters used for ML hyperparameter tuning. This table shows only parameters that were modified for the ML and DL hyperparameter-tuning tasks. The full list is shown in Table 4.2

Parameter	Value	Description
funEvals	Inf	
multiStart	2	
noise	Noise	
parNames	cfg$tunepars	
seedFun	123	
Time	List (maxTime = timebudget/60)	Convert to minutes
transformFun	cfg$transformations	
Types	cfg$type	
designControl	Replicates	Rinit
	Size	initSizeFactor * length(cfg$lower)
yImputation	List	
modelControl	funEvals	multFun * length(cfg$lower)
optimizerControl	funEvals	multFun * length(cfg$lower)

8.6 Tunability

The following analysis is based on the results from the spot run, which are stored in the data folder of this book. They can be loaded with the following command:

```
load("supplementary/ch08-CaseStudyI/ranger00001.RData")
```

Now the information generated with spot, which was stored in the result list as described in Sect. 8.5, is available in the R environment.

8.6.1 Progress

The function prepareProgressPlot generates a data frame that can be used to visualize the hyperparameter-tuning progress. The data frame can be passed to ggplot. Figure 8.6 visualizes the progress during the ranger hyperparameter-tuning process described in this study.

After 60 min, 582 ranger models were evaluated. Comparing the worst configuration that was observed during the HPT with the best, a 25.8442 % reduction was obtained. After the initial phase, which includes 20 evaluations, the smallest MMCE reads 0.179964. The dotted red line in Fig. 8.6 illustrates this result. The final best value reads 0.1712657, i.e., a reduction of the MMCE of 4.8333%. These values, in

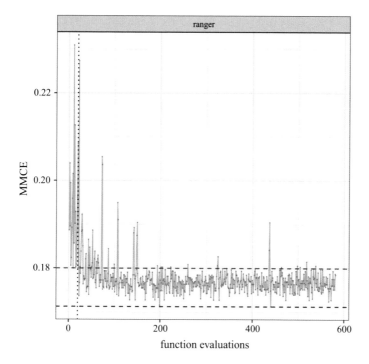

Fig. 8.6 Ranger: Hyperparameter-tuning progress. The *red* dashed line denotes the best value found by the initial design. The *blue* dashed line represents the best value from the whole run

combination with results shown in the progress plot (Fig. 8.6) indicate that a quick HPT run is able to improve the quality of the `ranger` method. It also indicates, that increased run times do not result in a significant improvement of the MMCE.

> **! Attention**
>
> These results do not replace a sound statistical comparison, they are only indicators, not final conclusions.

8.6.2   `evalParamCensus`: *Comparing Default and Tuned Parameters on Test Data*

As a comparison basis, an additional experiment for the `ranger` model where all hyperparameter values remain at the model's default settings and an additional experiment where the tuned hyperparameters are used, is performed. In these cases, a `timeout` for evaluation was not set. Since no search takes place, the overall run time for default values is anyways considerably lower than the run time of spot. The final comparison is based on the classification error as defined in Eq. (2.2). The motivation for this comparison is a consequence of the tunability definition; see Definition 2.26.

To understand the impact of tuning, the best solution obtained is evaluated for n repeats and compared with the performance (MMCE) of the default settings. A power analysis, as described in Sect. 5.6.5 is performed to estimate the number of repeats, n.

The corresponding values are shown in Table 8.9. The function evalParamCensus was used to perform this comparison. Results from the evaluations on the test data for the default and the tuned hyperparameter configurations are saved to the corresponding files.

Default and tuned results for the ranger model are available in the supplementary data folder as rangerDefaultEvaluation.RData and ranger00001Evaluation.RData, respectively.

> **Important:**

As explained in Sect. 8.4.4, no explicit validation set, $X^{(\text{val})}$, is defined during the HPT procedure. The response surface function $\psi^{(\text{test})}$ is optimized. But, since we can generate new data sets, (X, Y) randomly, the comparison is based on several, randomly generated samples.

Background: Additional Scores

The scores are stored as a matrix. Attributes are used to label the measures. In addition to mmce, the following measures are calculated for each repeat: accuracy, f1, logLoss, mae, precision, recall, and rmse. These results are stored in the corresponding RData files.

Hyperparameters of the default and the tuned configurations are shown in Table 8.9.

Table 8.9 Comparison of default and tuned hyperparameters of the "ranger" model. *r.u.f.* denotes respect.unordered.factors and *s.f* sample.fraction

Hyperparam.	num.trees	mtry	s.f	Replace	r.u.f	Min.	1st Qu.	Median	Mean	3rd Qu.	Max.
Default	8.966	4	1.0	2	1	0.1803	0.1879	0.1929	0.1913	0.1952	0.1998
Tuned	9.305	20.000	0.142	2.000	2.000	0.1737	0.1809	0.1872	0.1856	0.1889	0.1986
Tuned OCBA	9.305	20.000	0.142	2.000	2.000	0.1737	0.1809	0.1872	0.1856	0.1889	0.1986

Fig. 8.7 Comparison of ranger algorithms with default (D) and tuned (T) hyperparameters. Classification error (MMCE). Vertical lines mark quantiles (0.25, 0.5, 0.75) of the corresponding distribution. Numerical values are shown in Table 8.9

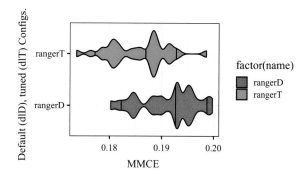

The corresponding R code for replicating the experiment is available in the `code` folder. The result files can be loaded and the violin plot of the obtained MMCE can be visualized as shown in Fig. 8.7. It can be seen that the tuned solutions provide a better MMCE on the holdout test data set $(X, Y)^{(\text{test})}$.

8.7   Analyzing the Random Forest Tuning Process

To analyze effects and interactions between hyperparameters of the `ranger` model as defined in Table 8.6, a simple regression tree as shown in Fig. 8.8 can be used.

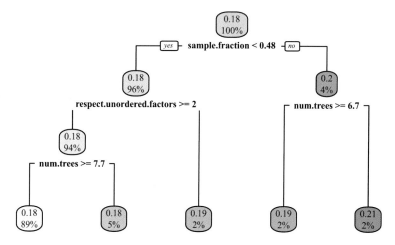

Fig. 8.8 Regression tree. Case study I. Ranger

The regression tree supports the observations, that hyperparameter values for sample.fraction, num.trees, and respect.unordered.factors, have the largest effect on the MMCE.

	sample.fraction	num.trees	respect.unordered.factors	mtry	replaRce
1	0.010297452	0.006007073	0.0015083938	0.0013354262	0.00015087878

Parallel plots visualize relations between hyperparameters. The SPOTMisc function ggparcoordPrepare provides an interface from the data frame result, which is returned from the function spot, to the function ggparcoord from the package GGally. The argument probs specifies the quantile probabilities for categorizing the result values. In Fig. 8.9, quantile probabilities are set to c(0.25, 0.5, 0.75). Specifying three values results in four categories with increasing performance, i.e., the first category (0–25%) contains poor results, the second and the third categories, 25–50 % and 50 to 75%, respectively, contain mediocre values, whereas the last category (75–100%) contains the best values.

In addition to labeling the best configurations, the worst configurations can also be labeled.

Results from the spot run can be passed to the function plotSenstivity, which generates a sensitivity plot as shown in Fig. 8.10. There are basically two types of sensitivity plots that can be generated with plotSenstivity: using the argument type = "best", the best hyperparameter configuration is used. Alternatively, using type = "agg", simulations are performed over the range of all hyperparameter settings. Note, the second option requires additional computations and depends on the simulation output, which is usually non-deterministic. Output from the second option is shown in Fig. 8.11.

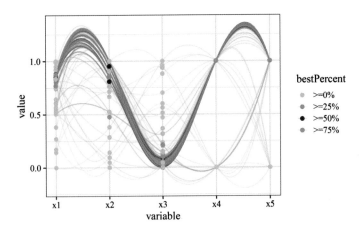

Fig. 8.9 Parallel plot of results from the ranger hyperparameter-tuning process. num.trees (x_1), mtry (x_2), sample.fraction (x_3), replace (x_4), and respect.unordered.factors (x_5) are shown. Best configurations in green

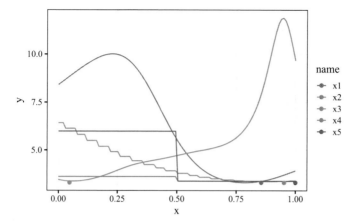

Fig. 8.10 Sensitivity plot (best). num.trees (x_1), mtry (x_2), sample.fraction (x_3), replace (x_4), and respect.unordered.factors (x_5) are shown

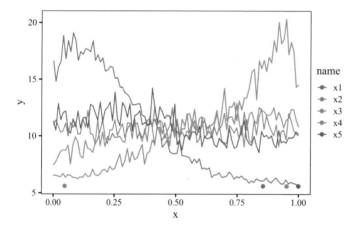

Fig. 8.11 Ranger: Sensitivity plot (aggregated). num.trees (x_1), mtry (x_2), sample.fraction (x_3), replace (x_4), and respect.unordered.factors (x_5) are shown

If the results from using the argument `type = "best"` and `type = "agg"` are qualitatively similar, only the plot based on `type = "best"` will be shown in the remainder of this book. Parallel plots will be treated in a similar manner. Source code for generating all plots is provided.

SPOT provides several tools for the analysis of interactions. Highly recommended is the use of contour plots as shown in Fig. 8.12.

Finally, a simple linear regression model can be fitted to the data. Based on the data from SPOT's `result` list, the hyperparameters `replace` and `respect.unordered.factors` are converted to `factors` and the R function `lm` is applied. The summary table is shown below.

Fig. 8.12 Surface plot: x_3 (sample.fraction) plotted against x_1 (numtrees)

```
##
## Call:
## lm(formula = y ~ ., data = df)
##
## Residuals:
##         Min         1Q      Median         3Q        Max
## -0.0122996 -0.0013971 -0.0000444  0.0014070  0.0162966
##
## Coefficients:
##                                  Estimate Std. Error t value Pr(>|t|)
## (Intercept)                     1.991e-01  1.198e-03 166.275   <2e-16 ***
## num.trees                      -2.220e-03  1.052e-04 -21.115   <2e-16 ***
## mtry                            3.907e-05  3.647e-05   1.071   0.2845
## sample.fraction                 1.716e-02  1.038e-03  16.533   <2e-16 ***
## replaceTRUE                     1.431e-03  5.580e-04   2.564   0.0106 *
## respect.unordered.factorsTRUE  -6.276e-03  6.378e-04  -9.840   <2e-16 ***
## ---
## Signif. codes:  0 '***' 0.001 '**' 0.01 '*' 0.05 '.' 0.1 ' ' 1
##
## Residual standard error: 0.0026 on 576 degrees of freedom
## Multiple R-squared:  0.7682, Adjusted R-squared:  0.7662
## F-statistic: 381.7 on 5 and 576 DF,  p-value: < 2.2e-16
```

Although this linear model requires a detailed investigation (a mispecification analysis is necessary), it also is in accordance with previous observations that hyperparameters sample.fraction, num.trees, and respect.unordered.factors have significant effects on the loss function.

Results indicate that sample.fraction is the dominating hyperparameter. Its setting has the largest impact on `ranger`'s performance. For example, the sensitivity plot Fig. 8.10 shows that small sample.fraction values improve the performance. The larger values clearly improve the performance. The regression tree analysis (see Fig. 8.8) supports this hypothesis, because sample.fraction is the root node of the

Table 8.10 Case study I: result analysis

p-value	Decision	Power	Cohen's d	Hedge's g	Severity
0	H0 rejected	0.9999941	1.0329001	1.0194859	$\Delta \leq 0.005$ are well supported

tree and values smaller than 0.48 are recommended. Furthermore, the regression tree analysis indicates that additional improvements can be obtained if the num.trees is greater equal 2. These observations are supported by the parallel plots and surface plots, too. The linear model can be interpreted in a similar manner.

8.8 Severity: Validating the Results

Now, let us proceed to analyze the statistical significance of the achieved performance improvement. The results from the pre-experimental runs indicate that the difference is $\bar{x} = 0.0057$. As this value is positive, for the moment, let us assume that the tuned solution is superior. The corresponding standard deviation is $s_d = 0.0045$. Based on Eq. 5.14, and with $\alpha = 0.05$, $\beta = 0.2$, and $\Delta = 0.005$ let us identify the required number of runs for the the full experiment using the getSampleSize function.

For a relevant difference of 0.005, approximately 10 runs per algorithm are required. Since, we evaluated for 30 repeats, we can now proceed to evaluate the severity and analyse the performance improvement achieved through tuning the parameters of the ranger.

The summary result statistics is presented in Table 8.10. The decision based on p-value is to reject the null hypothesis, i.e., the claim that the tuned parameter setup provides a significant performance improvement in terms of MMCE is supported. The effect size suggests that the difference is of larger magnitude. For the chosen $\Delta = 0.005$, the severity value is at 0.8 and thus it strongly supports the decision of rejecting the H_0. The severity plot is shown in Fig. 8.13. Severity shows that performance difference smaller than or equal to 0.005 are well supported.

8.9 Summary and Discussion

The analysis indicates that hyperparameter sample.fraction has the greatest effect on the algorithm's performance. The recommended value of sample.fraction is 0.1416, which is much smaller than of 1.

This case study demonstrates how functions from the R packages mlr and SPOT can be combined to perform a well-structured hyperparameter tuning and analysis. By specifying the time budget via maxTime, the user can systematically improve hyperparameter settings. Before applying ML algorithms such as RF to complex

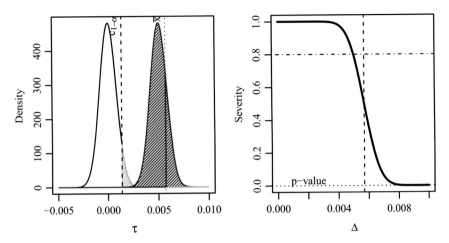

Fig. 8.13 Tuning Random Forest. Severity of rejecting H0 (red), power (blue), and error (gray). Left: the observed mean $\bar{x} = 0.0057$ is larger than the cut-off point $c_{1-\alpha} = 0.0014$ Right: The claim that the true difference is as large 0.005 are well supported by severity. However, any difference larger than 0.005 is not supported by severity

classification or regression problems, HPT is recommended. Wrong hyperparameter settings can be avoided. Insight into the behavior of ML algorithms can be obtained.

8.10 Program Code

Program Code

```
library("SPOT")
library("SPOTMisc")

target <- "age"
task.type <- "classif"
nobs <- 1e4
nfactors <- "high"
nnumericals <- "high"
cardinality <- "high"
data.seed <- 1
cachedir <- "oml.cache"

dfCensus <- getDataCensus(
  task.type = task.type,
  nobs = nobs,
  nfactors = nfactors,
```

```r
    nnumericals = nnumericals,
    cardinality = cardinality,
    data.seed = data.seed,
    cachedir = cachedir,
    target = target
)

model <- "ranger"
cfg <- getMlConfig(
  target = target,
  model = model,
  data = dfCensus,
  task.type = task.type,
  nobs = nobs,
  nfactors = nfactors,
  nnumericals = nnumericals,
  cardinality = cardinality,
  data.seed = data.seed,
  prop = 2 / 3
)

task <- getMlrTask(
  dataset = dfCensus,
  task.type = "classif",
  data.seed = 1
)

nFeatures <- sum(task$task.desc$n.feat)
cfg <- getModelConf(
  task.type = task.type,
  model = model,
  nFeatures = nFeatures
)

rsmpl <- getMlrResample(
  task = task,
  dataset = dfCensus,
  data.seed = 1,
  prop = 2 / 3
)

timebudget <- 60 ## secs
timeout <- timebudget / 20

cfg <- append(cfg, list(
  task = task,
  resample = rsmpl
))

objf <- getObjf(
  config = cfg,
  timeout = timeout
)
```

```
result <- spot(
  x = NULL,
  fun = objf,
  lower = cfg$lower,
  upper = cfg$upper,
  control = list(
    types = cfg$type,
    time = list(maxTime = timebudget / 60),
    noise = TRUE,
    seedFun = 123,
    designControl = list(
      replicates = 2,
      size = length(cfg$lower)
    ),
    replicates = 2,
    funEvals = Inf,
    optimizerControl = list(funEvals = 200 * length(cfg$lower)),
    multiStart = 2,
    parNames = cfg$tunepars,
    yImputation = list(
      handleNAsMethod = handleNAsMean,
      imputeCriteriaFuns = list(is.infinite, is.na, is.nan),
      penaltyImputation = 3
    )
  )
)
```

Chapter 9
Case Study II: Tuning of Gradient Boosting (xgboost)

Thomas Bartz-Beielstein, Sowmya Chandrasekaran, and Frederik Rehbach

Abstract This case study gives a hands-on description of Hyperparameter Tuning (HPT) methods discussed in this book. The Extreme Gradient Boosting (XGBoost) method and its implementation xgboost was chosen, because it is one of the most powerful methods in many Machine Learning (ML) tasks, especially when standard tabular data should be analyzed. This case study follows the same HPT pipeline as the first and third studies: after the data set is provided and pre-processed, the experimental design is set up. Next, the HPT experiments are performed. The R package SPOT is used as a "datascope" to analyze the results from the HPT runs from several perspectives: in addition to Classification and Regression Trees (CART), the analysis combines results from the surface, sensitivity, and parallel plots with a classical regression analysis. Severity is used to discuss the practical relevance of the results from an error-statistical point-of-view. The well-proven R package mlr is used as a uniform interface from the methods of the packages SPOT and SPOTMisc to the ML methods. The corresponding source code is explained in a comprehensible manner.

9.1 Introduction

This chapter considers the XGBoost algorithm which was detailed in Sect. 3.6. How to find suitable parameter values and bounds, and how to perform experiments w.r.t. the following nine XGBoost hyperparameters will be discussed: nrounds,

Supplementary Information The online version contains supplementary material available at https://doi.org/10.1007/978-981-19-5170-1_9.

T. Bartz-Beielstein (✉) · S. Chandrasekaran · F. Rehbach
Institute for Data Science, Engineering and Analytics, TH Köln, Cologne, Germany
e-mail: thomas.bartz-beielstein@th-koeln.de

S. Chandrasekaran
e-mail: sowmya.chandrasekaran@th-koeln.de

F. Rehbach
e-mail: frederik.rehbach@th-koeln.de

eta, lambda, alpha, subsample, colsample, gamma, maxdepthx, and minchild.

9.2 Data Description

The first step is identical to the step in the ranger example in Chap. 8, because the Census-Income (KDD) Data Set (CID) will be used.[1] So, the function getDataCensus is called with the parameters from Table 8.3 to get the CID data from Table 8.2. The complete data set, (X, Y) contains $n = 299,285$ observations with 41 features on demography and employment.

9.3 getMlConfig: Experimental Setup and Configuration of the Gradient Boosting Model

Again, a subset with $n = 1e4$ samples that defines the subset $(X, Y) \in (X, Y)$ is provided. The project setup is also similar to the setup described in Sect. 8.1. Therefore, only the differences will be shown. The full script is available in Sect. 9.10.

The function getMlConfig is called with the same arguments as in Chap. 8, with one exception: model is set to "xgboost". The function getMlConfig defines the ML task, the model configuration, and the data split (generation of the training and test data sets, i.e., $(X, Y)^{(\text{train})}$ and $(X, Y)^{(\text{test})}$.) To achieve this goal, the functions getMlTask, getModelConf, and getMlrResample are executed. As a result, the list cfg with 13 elements is available, see Table 9.1.

```
model <- "xgboost"
cfg <- getMlConfig(
   target = target,
   model = model,
   data = dfCensus,
   task.type = task.type,
   nobs = nobs,
   nfactors = nfactors,
   nnumericals = nnumericals,
   cardinality = cardinality,
   data.seed = data.seed,
   prop = 2 / 3
)
```

---

[1] The data from CID is historical. It includes wording or categories regarding people which do not represent or reflect any views of the authors and editors.

Table 9.1 Result from the function `getMlConfig`: the `cfg` list

Parameter	Type	Value
learner	chr	"classif.xgboost"
tunepars	chr [1:9]	"nrounds" "eta" "lambda" "alpha" ..., see Table 9.2
defaults	num [1, 1:9]	0 -1.74 0 -10 1 ...
lower	num [1:9]	0 -10 -10 -10 0.1 ...
upper	num [1:9]	11 0 10 10 1 1 10 15 7
type	chr [1:9]	"numeric" "numeric" "numeric" "numeric" ...
fixpars	list	eval_metric: chr "logloss" and number of threads
factorlevels	list()	
transformations:	List of 9	Transformation functions, see Table 9.2
dummy	logi	TRUE
relpars	list()	Empty list (no parameters are relative to each other)
task	List of 6	mlr task object
resample	List of 5	Resample information

9.3.1 `getMlrTask`: Problem Design and Definition of the Machine Learning Task

The problem design describes the target and task type, the number of observations, as well as the number of factorial, numerical, and cardinal variables. It was described in Sect. 8.3.2.1.

9.3.2 `getModelConf` Algorithm Design—Hyperparameters of the Models

The function `getModelConf`, which is called from `getMlConf`, computes an adequate XGBoost hyperparameter setting. Examples from literature shown in Table 3.6 in Sect. 3.6 will be used as a guideline. These values were modified as follows:

nrounds:
: An upper value (2^5), which is similar to the Random Forest (RF) configuration, was chosen. This value is smaller than the value used by Probst et al. (2019a), who used 5 000.

colsample_bytree:
: The lower value was chosen as 1/getTaskNFeats (task). This is a minor deviation from the settings used in Probst et al. (2019a). The reason for this modification

Table 9.2 XGBoost hyperparameter. N_{Feats} denotes the output from `getTaskNFeats(task)`

Name	Type	Default	Lower (mlr)	Upper (mlr)	Upper (SPOT)	Lower (SPOT)	Trans
nrounds	integer	–	1	Inf	0	5	2pow_round
eta	numeric	0.3	0	1	–10	0	2pow
lambda	numeric	1	0	Inf	–10	10	2pow
alpha	numeric	0	0	Inf	–10	10	2pow
subsample	numeric	1	0	1	0.1	1	id
colsample_bytree	numeric	1	0	1	$1/N_{Feats}$	1	id
gamma	numeric	0	0	Inf	–10	10	2pow
max_depth	integer	6	0	Inf	1	15	id
min_child_weight	numeric	1	0	Inf	0	7	2pow

gamma:

is simple: a lower value of zero makes no sense, because at least one feature should be chosen via `colsample`. A lower value of -10 was chosen. This value is smaller than the value chosen by Thomas et al. (2018). Accordingly, a larger upper value (10) than by Thomas et al. (2018) was selected.

Hyperparameter transformations are shown in the column `trans` in Table 9.2. These transformations are similar to the transformations used by Probst et al. (2019a) and Thomas et al. (2018) with one minor change: `trans_2pow_round` was applied to the hyperparameter `nrounds`.

The ML configuration list `cfg` contains information about the hyperparameters of the XGBoost model, see Table 9.2.

Background: XGBoost Hyperparameters

The complete list of XGBoost hyperparameters can also be shown using the function `getModelConf`. Note: the hyperparameter `colsample_bytree` is a relative hyperparameter, i.e., it depends on the number of features (`nFeatures`), see the discussion in Sect. 3.6. Hence, the value `nFeatures` must be determined before the hyperparameter bounds can be computed.

```
nFeatures <- sum(task$task.desc$n.feat)
modelCfg <- getModelConf(
  task.type = task.type,
  model = model,
  nFeatures = nFeatures
)
```

The list of hyperparameters is stored as the list element `tunepars`, see Table 9.2.

Furthermore, all factor features will be replaced with their dummy variables. Dummy variables are recommended for XGBoost: internally, a `model.matrix` is used and non-factor features will be left untouched and passed to the result. The seed can be set to improve reproducibility. Finally, these settings are compiled to the list `cfg`.

9.3.3 *getMlrResample: Training and Test Data*

The partition of the full data set is done as described in Sect. 8.3.2.3. `rsample` specifies a training data set, which contains 2/3 of the data and a testing data set with the remaining 1/3 of the data.

9.4 Objective Function (Model Performance)

Because the XGBoost method is more complex than RF, an increased computational budget is recommended, e.g., by choosing a budget for tuning of $6 \times 3{,}600$ s or six hours. The increased budget is used in the global study (Chap. 12). For the experiments performed in the current chapter, the budget was not increased.

Before the hyperparameter tuner is called, the objective function is defined: this function receives a configuration for a tuning experiment and returns an objective function to be tuned via `spot`. A detailed description of the objective function can be found in Sect. 8.4.4.

9.5 spot: Experimental Setup for the Hyperparameter Tuner

The R package `SPOT` is used to perform the actual tuning (optimization). Because the generic Sequential Parameter Optimization Toolbox (SPOT) setup was introduced in Sect. 4.5, this section highlights the modifications of the generic setup that were made for the `xgboost` hyperparameter tuning experiments.

The third step of the hyperparameter tuning pipeline as shown in Fig. 8.5 starts the SPOT hyperparameter tuner.

```
result <- spot(
  x = NULL,
  fun = objf,
  lower = cfg$lower,
  upper = cfg$upper,
  control = list(
```

```
types = cfg$type,
time = list(maxTime = timebudget / 60),
noise = TRUE,
OCBA = TRUE,
OCBABudget = 3,
seedFun = 123,
designControl = list(
  replicates = Rinit,
  size = initSizeFactor * length(cfg$lower)
),
replicates = 2,
funEvals = Inf,
modelControl = list(
  target = "ei",
  useLambda = TRUE,
  reinterpolate = FALSE
),
optimizerControl = list(funEvals = 200 * length(cfg$lower)),
multiStart = 2,
parNames = cfg$tunepars,
yImputation = list(
  handleNAsMethod = handleNAsMean,
  imputeCriteriaFuns = list(is.infinite, is.na, is.nan),
  penaltyImputation = 3
)
)
)
```

The result is written to a file and can be accessed via

```
load("supplementary/ch09-CaseStudyII/xgboost00001.RData")
```

The full R code for running this case study is shown in the Appendix (Sect. 9.10).

9.6 Tunability

9.6.1 Progress

The function prepareProgressPlot generates a data frame that can be used to visualize the hyperparameter tuning progress. The data frame can be passed to ggplot. Figure 9.1 visualizes the progress during the ranger hyperparameter tuning process during the spot tuning procedure.

After 60 min, 157 xgboost models were evaluated. Comparing the worst configuration that was observed during the HPT with the best, a 66.3743% reduction was obtained. After the initial phase, which includes 18 evaluations, the smallest Mean Mis-Classification Error (MMCE) reads 0.1793641. The dotted red line in Fig. 8.6 illustrates this result. The final best value reads 0.1724655, i.e., a reduction

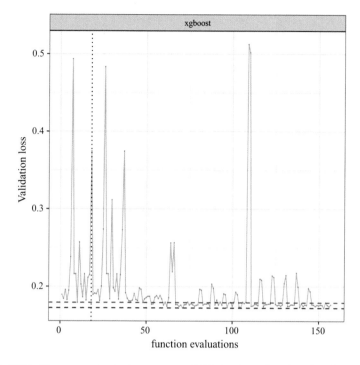

Fig. 9.1 XGB: Hyperparameter tuning progress. Validation loss plotted against the number of function evaluations, i.e., the number of evaluated XGBoost models. The red dashed line denotes the best value found by the initial designs to show the hyperparameter tuning progress. The blue dashed line represents the best value from the whole run

of the MMCE of 3.8462%. These values, in combination with results shown in the progress plot (Fig. 8.6), indicate that a quick HPT run is able to improve the quality of the xgboost method. It also indicates that increased run times do not result in a significant improvement of the MMCE.

! Attention

These results do not replace a sound statistical comparison, they are only indicators, not final conclusions.

Table 9.3 Comparison of the default and tuned hyperparameters of the XGBoost method. `colsample` denotes `colsample_bytree`. Table shows transformed values. Note: the `alpha` and `gamma` values are identical. They are computed as 2^{-10}, which is the lower bound value, because the theoretical default value 0 is infeasible. See also Table 3.8

Method	nrounds	eta	lambda	alpha	subsample	colsample	gamma	max_depth	min_child_weight
Default	1	0.3	1	0.001	1	1	0.001	6	1
Tuned	1873	0.058	155.3	2.46	0.992	0.408	0.004	13	1.83

Fig. 9.2 Comparison of XGBoost methods with default (D) and tuned (T) hyperparameters. Classification error (MMCE) plotted on the horizontal axis. Vertical lines in the violin figures mark quantiles (0.25, 0.5, 0.75) of the corresponding distribution. Numerical values are shown in Table 9.3

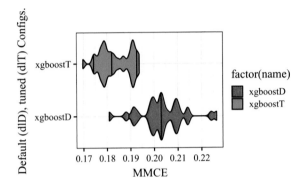

9.6.2 *evalParamCensus: Comparing Default and Tuned Parameters on Test Data*

As a baseline for comparison, XGBoost was run with default hyperparameter values. The corresponding R code for replicating the experiment is available in the `code` folder. The best (minimum MMCE) result from thirty repeats is reported. The corresponding values are shown in Table 9.3. The function `evalParamCensus` was used to perform this comparison. By specifying the ML model, e.g., `"xgboost"` and the `runNr`, the function `evalParamCensus` was called.

The result files can be loaded and the violin plot of the obtained MMCE can be visualized (Fig. 9.2). It can be seen that the tuned solutions provide a better MMCE. Default and tuned results for the `ranger` model are available as `rangerDefault Evaluation.RData` and `xgboost00001Evaluation.RData`, respectively.

The scores are stored as a `matrix`. Attributes are used to label the measures. The following measures are calculated for each hyperparameter setting: `accuracy`, `ce`, `f1`, `logLoss`, `mae`, `precision`, `recall`, and `rmse`. The comparison is based on the MMCE that was defined in Eq. (2.2). Hyperparameters of the default and the tuned configurations are shown in Table 9.3. The full procedure, i.e., starting from scratch, to obtain the default `xgboost` hyperparameters is shown in Sect. 9.10.

Next, the hyperparameters of the tuned `xgboost` methods are shown.

9.7   Analyzing the Gradient Boosting Tuning Process

The analysis and the visualizations are based on the transformed values.

To analyze effects and interactions between hyperparameters of the xgboost Model, a simple regression tree as shown in Fig. 9.3 and Fig. 9.4 can be used.

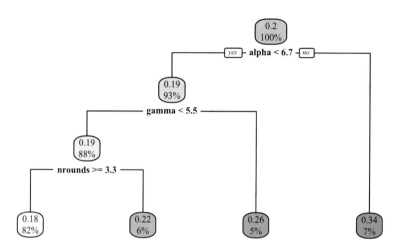

Fig. 9.3   Regression tree. Case study II. XGBoost

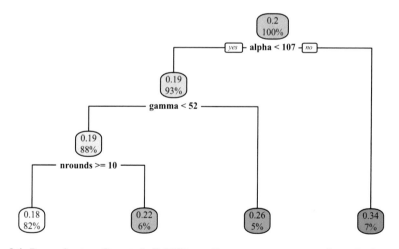

Fig. 9.4   Regression tree. Case study II. XGBoost. Hyperparameters are transformed values

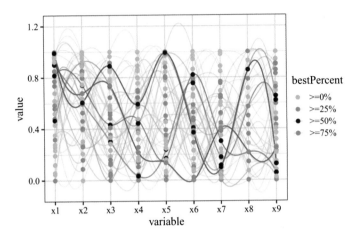

Fig. 9.5 Best configurations in green

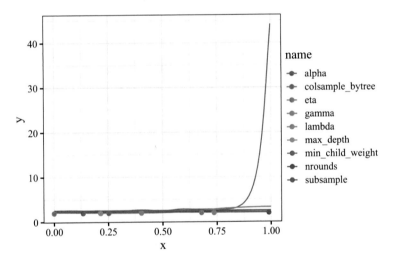

Fig. 9.6 Sensitivity plot (best). Too large alpha values result in poor results

Same tree with the transformed values:

The regression tree supports the observations that hyperparameter values for alpha, lambda, gamma, and nrounds have the largest effect on the MMCE.

	alpha	lambda	gamma	nrounds	subsample	eta	colsample_bytree
1	0.23112227	0.04431784	0.04039483	0.014028719	0.012203015	0.009397272	0.0028057437

alpha is the most relevant hyperparameter.

To perform a sensitivity analysis, parallel and sensitivity plots can be used (Figs. 9.5 and 9.6).

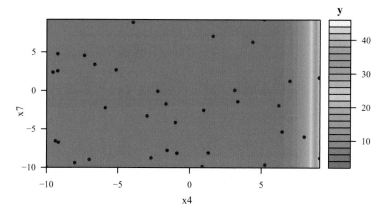

Fig. 9.7 Surface plot: x_3 plotted against x_1. This surface plot indicates that `alpha` has a large effect. Too large alpha values result in poor results

Results from the `spot` run can be passed to the function `plotSenstivity`, which generates a sensitivity plot as shown in Fig. 8.10. Sensitivity plots were introduced in Sect. 8.6. Contour plots are shown in Fig. 9.7.

Finally, a simple linear regression model can be fitted to the data. Based on the data from SPOT's `result` list, the summary is shown below.

```
##
## Call:
## lm(formula = y ~ ., data = df)
##
## Residuals:
##       Min        1Q    Median        3Q       Max
## -0.073397 -0.015307 -0.008367  0.001629  0.223535
##
## Coefficients:
##                    Estimate Std. Error t value Pr(>|t|)
## (Intercept)       0.2561669  0.0184896  13.855  < 2e-16 ***
## nrounds          -0.0032016  0.0012827  -2.496  0.01366 *
## eta              -0.0003031  0.0016994  -0.178  0.85870
## lambda            0.0021936  0.0007939   2.763  0.00646 **
## alpha             0.0072423  0.0007848   9.228 2.77e-16 ***
## subsample        -0.1033194  0.0137081  -7.537 4.55e-12 ***
## colsample_bytree  0.0050479  0.0132658   0.381  0.70411
## gamma             0.0010887  0.0009034   1.205  0.23007
## max_depth         0.0023106  0.0010527   2.195  0.02974 *
## min_child_weight  0.0082803  0.0025515   3.245  0.00145 **
## ---
## Signif. codes:  0 '***' 0.001 '**' 0.01 '*' 0.05 '.' 0.1 ' ' 1
##
## Residual standard error: 0.04047 on 147 degrees of freedom
## Multiple R-squared:  0.5222, Adjusted R-squared:  0.493
## F-statistic: 17.85 on 9 and 147 DF,  p-value: < 2.2e-16
```

Although this linear model requires a detailed investigation (a misspecification analysis is necessary) it also is in accordance with previous observations that hyperparameters `alpha`, `lambda`, `gamma`, `nrounds` have significant effects on the loss function.

9.8 Severity: Validating the Results

Now, we utilize hypothesis testing and severity to analyze the statistical significance of the achieved performance improvement. Considering the results from the preexperimental runs, the difference is $\bar{x} = 0.0199$. Since this value is positive, for the moment, let us assume that the tuned solution is superior. The corresponding standard deviation is $s_d = 0.0081$. Based on Eq. 5.14, and with $\alpha = 0.05$, $\beta = 0.2$, and $\Delta = 0.01$, let us identify the required number of runs for the full experiment using the `getSampleSize()` function.

Table 9.4 Case study II: result analysis

p-value	Decision	power	Cohen's d	Hedge's g	Severity
0	H0 rejected	0.9999999	2.3978067	2.3666664	$\Delta \leq 0.015$ are well supported

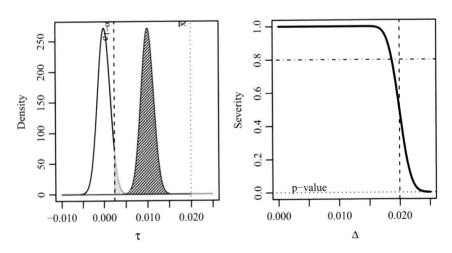

Fig. 9.8 Tuning XGB. Severity of rejecting H0 (red), power (blue), and error (gray). Left: The observed mean $\bar{x} = 0.0199$ is larger than the cut-off point $c_{1-\alpha} = 0.0024$ Right: The claim that the true difference is as large or larger than 0.01 is well supported by severity. However, any difference larger than 0.015 is not supported by severity

For a relevant difference of 0.01, approximately 8 runs per algorithm are required. Hence, we can proceed to evaluate the severity and analyze the performance improvement achieved through tuning the parameters of the xgboost.

The summary result statistics is presented in Table 9.4. The decision based on p-value is to reject the null hypothesis, i.e., the claim that the tuned parameter setup provides a significant performance improvement in terms of MMCE is supported. The effect size suggests that the difference is of a larger magnitude. For the chosen $\Delta = 0.01$, the severity value is at 1 and thus it strongly supports the decision of rejecting the H_0. The severity plot is shown in Fig. 9.8. Severity shows that performance differences smaller than 0.015 are well supported.

9.9 Summary and Discussion

The analysis indicates that hyperparameter alpha has the greatest effect on the algorithm's performance. The recommended value of alpha is 7.2791, which is much larger than the default value.

This case study demonstrates how functions from the R packages `mlr` and SPOT can be combined to perform a well-structured hyperparameter tuning and analysis. By specifying the time budget via `maxTime`, the user can systematically improve hyperparameter settings. Before applying ML algorithms such as XGBoost to complex classification or regression problems, HPT is recommended. Wrong hyperparameter settings can be avoided. Insight into the behavior of ML algorithms can be obtained.

9.10 Program Code

Program Code

```
target <- "age"
task.type <- "classif"
nobs <- 1e4
nfactors <- "high"
nnumericals <- "high"
cardinality <- "high"
data.seed <- 1
cachedir <- "oml.cache"

dfCensus <- getDataCensus(
  task.type = task.type,
  nobs = nobs,
  nfactors = nfactors,
  nnumericals = nnumericals,
```

```
    cardinality = cardinality,
    data.seed = data.seed,
    cachedir = cachedir,
    target = target
)
task <- getMlrTask(
    dataset = dfCensus,
    task.type = "classif",
    data.seed = 1
)

model <- "xgboost"
cfg <- getMlConfig(
    target = target,
    model = model,
    data = dfCensus,
    task.type = task.type,
    nobs = nobs,
    nfactors = nfactors,
    nnumericals = nnumericals,
    cardinality = cardinality,
    data.seed = data.seed,
    prop = 2 / 3
)
transformX(cfg$defaults, cfg$transformations)


##        [,1] [,2] [,3]              [,4] [,5] [,6]              [,7] [,8] [,9]
## [1,]      1  0.3    1 0.0009765625      1    1 0.0009765625        6    1
```

Chapter 10
Case Study III: Tuning of Deep Neural Networks

Thomas Bartz-Beielstein, Sowmya Chandrasekaran, and Frederik Rehbach

Abstract A surrogate model based Hyperparameter Tuning (HPT) approach for Deep Learning (DL) is presented. This chapter demonstrates how the architecture-level parameters (hyperparameters) of Deep Neural Networks (DNNs) that were implemented in `keras/tensorflow` can be optimized. The implementation of the tuning procedure is 100% accessible from R, the software environment for statistical computing. How the software packages (`keras`, `tensorflow`, and `SPOT`) can be combined in a very efficient and effective manner will be exemplified in this chapter. The hyperparameters of a standard DNN are tuned. The performances of the six Machine Learning (ML) methods discussed in this book are compared to the results from the DNN. This study provides valuable insights in the tunability of several methods, which is of great importance for the practitioner.

10.1 Introduction

The DNN hyperparameter study described in this chapter uses the same data and the same HPT process as the ML studies in Chaps. 8 and 9. Section 10.2 describes the data preprocessing. Section 10.3 explains the experimental setup and the configuration of the DL models. The objective function is defined in Sect. 10.4. The hyperparameter tuner, `spot`, is described in Sect. 10.5. Based on this setup, experimental results are analyzed: After discussing tunability based on the HPT progress in Sect. 10.6, default, λ_0 and tuned hyperparameters, $\lambda^\star$ are compared in Sect. 10.6.2.

Supplementary Information The online version contains supplementary material available at https://doi.org/10.1007/978-981-19-5170-1_10.

T. Bartz-Beielstein (✉) · S. Chandrasekaran · F. Rehbach
Institute for Data Science, Engineering and Analytics, TH Köln, Gummersbach, Germany
e-mail: thomas.bartz-beielstein@th-koeln.de

S. Chandrasekaran
e-mail: sowmya.chandrasekaran@th-koeln.de

F. Rehbach
e-mail: frederik.rehbach@th-koeln.de

© The Author(s) 2023
E. Bartz et al. (eds.), *Hyperparameter Tuning for Machine and Deep Learning with R*,
https://doi.org/10.1007/978-981-19-5170-1_10

235

Table 10.1 Deep-learning hyperparameter pipeline

Step	Description	Function	Result	Details
1	Get data	getDataCensus	dfCensus	Data frame
2.1	Split data into training, validation, and test data	getGenericTrainValTestData	Data	Partitioned data
2.2	Spec	genericDataPrep	specList	List with the following data
2.3.1	keras configuration	getKerasConf	kerasConf	Configuration list for keras
2.3.2	Model configuration	getModelConf	cfg	Model
3	Hyperparameter tuning	spot	Result	Result list
4	Evaluate on test data	evalParamCensus	Score	Metrics

The DL tuning process is analyzed in Sect. 10.7. Results are validated using severity in Sect. 10.8. A summary in Sect. 10.9 concludes this chapter. The DL hyperparameter tuning pipeline, that was used for the experiments, is summarized in Table 10.1 and illustrated in Fig. 10.1. The first sections in this chapter highlight the most important steps of this pipeline. The program code for performing the experiments is shown in Sect. 10.10.

keras is TensorFlow (TF)'s high-level Application Programming Interface (API) designed with a focus on enabling fast experimentation. TF is an open source software library for numerical computations with data flow graphs (Abadi et al. 2016). Mathematical operations are represented as nodes in the graph, and the graph edges represent the multidimensional arrays of data (tensors) (O'Malley et al. 2019). The full TF API can be accessed via the tensorflow package from within the R software environment for statistical computing and graphics (R).

The Appendix contains information on how to set up the required PYTHON software environment for performing HPT with keras, SPOT, and SPOTMisc. Source code for performing the experiments will included in the R package SPOTMisc. Further information is published on https://www.spotseven.de and with some delay on Comprehensive R Archive Network (CRAN) (https://cran.r-project.org/package= SPOT). This delay is caused by an intensive code check, which is performed by the CRAN team. It guarantees high-quality open source software and is an important feature for providing reliable software that is not just *a flash in the pan*.

Fig. 10.1 Overview. The
HPT pipeline introduced in
this chapter comprehends the
following steps: After the
data acquisition
(getDataCensus), the
data is split into training,
validation, and test sets.
These data sets are processed
via the function
genericDataPrep.
keras is configured via the
function getKerasConf.
The hyperparameter tuner
spot is called and finally,
the results are evaluated
(evalParamCensus)

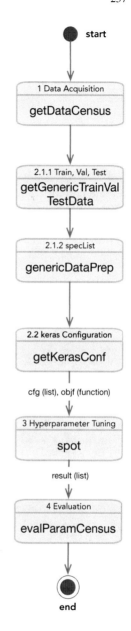

10.2   Data Description

Identically to the ML case studies, the DL case study presented in this chapter uses the Census-Income (KDD) Data Set (CID), which is made available, for example, via the University of California, Irvine (UCI) Machine Learning Repository.[1,2]

10.2.1   *getDataCensus: Getting the Data from OpenML*

Before training the DNN, the data is preprocessed by reshaping it into the shape the DNN can process. The function getDataCensus is used to get the Open Machine Learning (OpenML) data (from cache or from server). The same options as in the previous ML studies will be used, i.e., the parameter settings from Table 8.3 will be used.

```
target <- "age"
task.type <- "classif"
nobs <- 1e4
nfactors <- "high"
nnumericals <- "high"
cardinality <- "high"
data.seed <- 1
cachedir <- "oml.cache"
prop <- 2 / 3
dfCensus <- getDataCensus(
  task.type = task.type,
  nobs = nobs,
  nfactors = nfactors,
  nnumericals = nnumericals,
  cardinality = cardinality,
  data.seed = data.seed,
  cachedir = cachedir,
  target = target
)
```

10.2.2   *getGenericTrainValTestData: Split Data in Train, Validation, and Test Data*

The data frame dfCensus, $(X, Y) \subset (\mathcal{X}, \mathcal{Y})$, with 10 000 observations of 23 variables, is available. Based on prop, the data is split into training, validation, and test

---

[1] https://archive.ics.uci.edu/ml/datasets/Census-Income+(KDD).

[2] The data from CID is historical. It includes wording or categories regarding people which do not represent or reflect any views of the authors and editors.

data sets, $(X, Y)^{(\text{train})}$, $(X, Y)^{(\text{val})}$, and $(X, Y)^{(\text{test})}$, respectively. If `prop = 2/3`, the training data set has 4 444 observations, the validation data set has 2 222 observations, and the test data set the remaining 3 334 observations.

```
data <- getGenericTrainValTestData(dfGeneric = dfCensus, prop = prop)
```

10.2.3   genericDataPrep: *Spec*

The third step of the data preprocessing generates a `specList`.

```
batch_size <- 32
specList <- genericDataPrep(data = data, batch_size = batch_size)
```

The function `genericDataPrep` works as described in Sects. 10.2.3.1–10.2.3.5.

10.2.3.1   The Iterator: Data Frame to Data Set

The helper function `df_to_dataset`[3] converts the data frame `dfCensus` into a data set. This procedure enables processing of very large Comma Separated Values (CSV) files (so large that they do not fit into memory). The elements of the training data sets are randomly shuffled. Finally, consecutive elements of this data set are combined into batches.

Applying the function `df_to_dataset` generates a list of tensors. Each tensor represents a single column. The most significant difference to R's data frames is that a TF data set is an iterator.

```
train_ds_generic <-
  df_to_dataset(data$trainGeneric, batch_size = batch_size)
val_ds_generic <-
  df_to_dataset(data$valGeneric, shuffle = FALSE, batch_size = batch_size)
```

Background: Iterators

Each time an iterator is called it will yield a different batch of rows from the data set. The iterator function `iter_next` can be called as follows, so that batches are shown.

```
train_ds %>%
  reticulate::as_iterator() %>%
  reticulate::iter_next()
```

---

[3] https://tensorflow.rstudio.com/tutorials/advanced/structured/classify/.

The data set `train_ds_generic` returns a list of column names (from the data frame) that map to column values from rows in the data frame.

10.2.3.2   The feature_spec Object: Specifying the Target

TF has built-in methods to perform common input conversions.[4] The powerful `feature_column` system will be accessed via the user-friendly, high-level interface called `feature_spec`. While working with structured data, e.g., CSV data, column transformations and representations can be initialized and specified. A practical benefit of implementing data preprocessing within model $\mathcal{A}$ is that when $\mathcal{A}$ is exported, the preprocessing is already included. In this case, new data can be passed directly to $\mathcal{A}$.

> **! Attention: Keras Preprocessing Layers**
>
> `keras` and `tensorflow` are under constant development. The current implementation in `SPOTMisc` classifies structured data with feature columns. The corresponding TF module was designed for the use with TF version 1 estimators. It does fall under compatibility guarantees.[5] The newly developed `keras` module uses "preprocessing layers" for building `keras`-native input processing pipelines. Future versions of `SPOTMisc` will be based on preprocessing layers. However, because the underlying ideas of both preprocessing layers are similar (TF provides a migration guide[6]), the most important preprocessing steps will be presented next.

First the `spec` object `specGeneric` is defined. The response variable, here: `target`, can be specified using a formula, see Chambers and Hastie (1992) and the R function `formula`.

```
specGeneric <- feature_spec(dataset = train_ds_generic, target ~ .)
```

10.2.3.3   Adding Steps to the feature_spec Object

The CID data set contains a variety of data types. These mixed data types are converted to a fixed-length vector for the DL model to process. Based on their *feature type*, their *data type* or *level*, the columns will be treated differently. After creating the `feature_spec` object the step functions from Table 10.2 can be used to

---

[4] https://tensorflow.rstudio.com/tutorials/beginners/load/load_csv/.

[5] https://www.tensorflow.org/tutorials/structured_data/feature_columns?authuser=0.

[6] https://www.tensorflow.org/guide/migrate/migrating_feature_columns.

Table 10.2 Steps: data transformations depending on the data type

Step functions	Description
step_numeric_column	Numeric variables
step_categorical_with_vocabulary_list	Categorical variables with a fixed vocabulary
step_categorical_column_with_hash_bucket	Categorical variables using the hash trick
step_categorical_column_with_identity	Categorical variables stored as integers
step_categorical_column_with_vocabulary_file	Vocabulary stored in a file

Table 10.3 Description of the CID feature and data types that are used in the data set $(X, Y) \subset (\mathcal{X}, \mathcal{Y})$

Column	Feature type	Data type/levels
capital_gains	Num	Double
capital_losses	Num	Double
dividends_from_stocks	Num	Double
wage_per_hour	Num	Double
weeks_worked_in_year	Num	Integer
class_of_worker	Factor	9
industry_code	Factor	51
occupation_code	Factor	47
Education	Factor	17
marital_status	Factor	7
major_industry_code	Factor	24
major_occupation_code	Factor	15
Race	Factor	5
hispanic_origin	Factor	10
Sex	Factor	2
tax_filer_status	Factor	6
detailed_household_and_family_stat	Factor	29
detailed_household_summary_in_household	Factor	8
country_of_birth_self	Factor	42
Citizenship	Factor	5
income_class	Factor	2
Target	Factor	2

add further `steps`. Depending on the data type, the step functions specify the data transformations. Table 10.3 shows these types.

The R package `tfdatasets` provides selectors to select certain variable types and ranges, e.g., `all_numeric` to select all numeric variables, `all_nominal` to select all characters, or `has_type("float32")` to select variables based on their TF variable type. Based on the feature and data type shown in Table 10.3, the data transformations from Table 10.2 are applied. We will consider feature specs for continuous and catergorical data separately.

10.2.3.4   Feature Spec: Continuous Data

For continuous data, i.e., numerical variables, the function `step_numeric_column` will be used and all numeric variables will be normalized (scaled). The R package `tfdataset` provides the scaler function `scaler_min_max`, which uses the minimum and maximum of the numeric variable and the function `scaler_standard`, which uses the mean and the standard deviation.

10.2.3.5   Feature Spec: Categorical Data

The DNN model $\mathcal{A}$ cannot directly process categorical (nominal) data—they must be transformed so that they can be represented as numbers. The representation of categorical variables as a set of one-hot encoded columns is widely used in practice (Chollet and Allaire 2018). There are basically two options for specifying the kind of numeric representation used for categorical variables: indicator columns or embedding columns.

Background: Embedding

Suppose instead of having a factor with a few levels (e.g., three categorical features such as `red`, `green`, or `blue`), there are hundreds or even more levels. As the number of levels grows very large, it becomes unfeasible to train a DNN using one-hot encodings. In this situation, *embedding* should be used: instead of representing the data as a very large one-hot vector, the data can be stored as a low-dimensional vector of real numbers. Note, the size of the embedding is a parameter that must be tuned (Abadi et al. 2015).

The implementation in SPOTMisc uses two steps: first, based on the number of `levels`, i.e., the value of the parameter `minLevelSizeEmbedding` in the following code, the set of columns where embedding should be used, is determined. Then, either the function `step_indicator_column` or the function `step_embedding_column` is applied.

```
minLevelSizeEmbedding <- 100
embeddingDim <- floor(log(minLevelSizeEmbedding))
df <- data$trainGeneric
df <- df[-which(names(df) == "target")]
embeddingVars <-
```

```
  names(df %>%
    mutate_if(is.character, factor) %>%
    select_if(~ is.factor(.) & nlevels(.) > minLevelSizeEmbedding))
noEmbeddingVars <-
  names(df %>%
    mutate_if(is.character, factor) %>%
    select_if(~ is.factor(.) & nlevels(.) <= minLevelSizeEmbedding))
specGeneric <- specGeneric %>%
  step_numeric_column(all_numeric(),
    normalizer_fn = scaler_standard()
  ) %>%
  step_categorical_column_with_vocabulary_list(all_nominal()) %>%
  step_indicator_column(matches(noEmbeddingVars)) %>%
  step_embedding_column(matches(embeddingVars), dimension = embeddingDim)
```

After adding a `step` we need to `fit` the `specGeneric` object:

```
specGeneric_prep <- fit(specGeneric)
```

Finally, the following data structures are available:

1. `train_ds_generic` (batched, based on 4444 samples)
2. `val_ds_generic`, (batched, based on 2222 samples)
3. `specGeneric_prep` and
4. `testGeneric` (the remaining 3334 samples).

These data are returned as the list `specList` from the function `genericData Prep`.

```
specList <- genericDataPrep(data = data, batch_size = batch_size)
```

Dense features prepared with TF's feature columns mechanism can be listed. There are 22 dense features that will be passed to the DNN.

```
names(specList$specGeneric_prep$dense_features())
##  [1] "wage_per_hour"
##  [2] "capital_gains"
##  [3] "capital_losses"
##  [4] "divdends_from_stocks"
##  [5] "num_persons_worked_for_employer"
##  [6] "weeks_worked_in_year"
##  [7] "indicator_class_of_worker"
##  [8] "indicator_industry_code"
##  [9] "indicator_major_industry_code"
## [10] "indicator_occupation_code"
## [11] "indicator_major_occupation_code"
## [12] "indicator_education"
## [13] "indicator_marital_status"
## [14] "indicator_race"
## [15] "indicator_hispanic_origin"
```

```
## [16] "indicator_sex"
## [17] "indicator_tax_filer_status"
## [18] "indicator_detailed_household_and_family_stat"
## [19] "indicator_detailed_household_summary_in_household"
## [20] "indicator_country_of_birth_self"
## [21] "indicator_citizenship"
## [22] "indicator_income_class"
```

10.3 Experimental Setup and Configuration of the Deep Learning Models

10.3.1 getKerasConf: keras and Tensorflow Configuration

Setting up the keras configuration from within SPOTMisc is a simple step: the function getKerasConf is called. The function getKerasConf passes additional parameters to the keras function, e.g.,

activation: Activation function in the last Neural Network (NN) layer. Default: "sigmoid".

active: Vector of active variables, e.g., c(1,10) specifies that only the first and tenth variable will be considered by spot. This mechanism allows the shrinking the full set of tunable parameters, say λ, to a smaller set, $\lambda^{(-)}$, if the user wants to investigate the tunability (or the effect) of one or only a few hyperparameters.

callbacks: List of callbacks to be called during training. Default: list().

clearSession: Whether to call k_clear_session or not at the end of keras modeling. Default: FALSE.

encoding: Encoding used during data preparation. Default: "oneHot".

loss: Loss function, $\mathcal{L}$, for the compile from the package keras. For example Binary Cross Entropy (BCE) loss as defined in Eq. (2.3).
Default: "loss_binary_crossentropy".

metrics: Metrics function for compile. Default: "binary_accuracy".

model: Model, $\mathcal{A}$, as specified via getModelConf. Default: "dl". Forthcoming versions of SPOTMisc will provide additional DNN model types, e.g., Convolutional Neural Networks (CNNs).

nClasses:
Number of classes in (multi-class) classification. Specifies the number of units in the last layer (before softmax). Default: 1 (binary classification).

resDummy:
If TRUE, generate dummy (mock up) result for testing. If FALSE, run keras and tensorflow evaluations. Default: FALSE.

returnValue:
Return value. Can be one of "trainingLoss", "negTrainingAccuracy", "validationLoss", "negValidationAccuracy", "testLoss", or "negTestAccuracy".

returnObject:
Return object. Can be one of "evaluation", "model", "pred". Default: "evaluation".

shuffle:
Logical (whether to shuffle the training data $(X, Y)^{(train)}$ before each epoch) or string (for "batch"). Used in the function df_to_dataset. "batch" is a special option for dealing with the limitations of the Hierarchical Data Format (HDF) version 5 data. It shuffles in batch-sized chunks. Default: FALSE.

testData:
Test data, $(X, Y)^{(test)}$, on which to evaluate the loss, $\mathcal{L}$, and any model metrics, $\psi^{(test)}$ at the end of the optimization using the function evaluate.

tfDevice:
Tensorflow device. CPU/GPU allocation. Passed to tensorflow via tf$device(kerasConf $tfDevice). Default: "/cpu:0" (use CPU only).

trainData:
Training data, $(X, Y)^{(train)}$, on which to evaluate the loss and any model metrics at the end of each epoch.

validationData:
Validation data, $(X, Y)^{(val)}$, on which to evaluate the loss $\psi^{(val)}$ and any model metrics at the end of each epoch.

validation_split:
Float between 0 and 1. Fraction of the training data $(X, Y)^{(train)}$ to be used internally by $\mathcal{A}$ as validation data $(X, Y)^{(valtrain)}$. $\mathcal{A}$ will set apart this fraction of the training data, will not train on it, and will evaluate the loss and any model metrics on $(X, Y)^{(valtrain)}$ at the end of each epoch. $(X, Y)^{(valtrain)}$ is selected from the last samples in the $(X, Y)^{(train)}$ data provided, before shuffling. Default: 0.2.

verbose:
Verbosity mode (0 = silent, 1 = progress bar, 2 = one line per epoch). Default: 0.

The default settings are useful for the binary classification task analyzed in this chapter. Only the parameter kerasConf$clearSession is set to TRUE and kerasConf$verbose is set to 0.

```
kerasConf <- getKerasConf()
kerasConf$clearSession <- TRUE
kerasConf$verbose <- 0
```

10.3.2 getModelConf: DL Hyperparameters

```
cfg <- getModelConf(model = "dl")
```

If the default values from the function getKerasConf are used, the vector of hyperparameter λ contains the following elements: the dropout rates (dropout rates of the layers will be tuned individually), the number of units (the number of single outputs from a single layer), the learning rate (controls how much to change the DNN model in response to the estimated error each time the model weights are updated), the number of training epochs (a training epoch is one forward and backward pass of a complete data set), the optimizer for the inner loop, O_{inner}, and its parameters (i.e., β_1, β_2 as well as ϵ) and the number of layers. These hyperparameters and their ranges are listed in Table 10.4.

Table 10.4 The hyperparameters, λ, for the DNN, which implements a fully connected network

Variable	Hyperparameter	Type	Default	Lower bound	Upper bound
x_1	dropout: first layer dropout rate	Numeric	0	0	0.4
x_2	dropoutfact: dropout multiplier	Numeric	0	0	0.5
x_3	units: units per first layer	Integer	32	1	32
x_4	unitsfact: units multiplier	Numeric	0.2	0.25	1
x_5	learning_rate: learning rate for the optimizer	Numeric	$1e-3$	$1e-6$	$1e-2$
x_6	epochs inner loop O_{inner} number of training epochs	Integer	16	8	128
x_7	beta_1	Numeric	0.9	0.9	0.99
x_8	beta_2	Numeric	0.999	0.999	0.9999
x_9	layers	Integer	1	1	4
x_{10}	epsilon	Numeric	$1e-7$	$1e-9$	$1e-8$
x_{11}	optimizer	Factor	5	1	7

Table 10.5 Optimizers that can be selected via hyperparameter x_{11}. Default optimizer O_{inner} is adam. The function `selectKerasOptimizer` from the `SPOTMisc` implements the selection. The corresponding R functions have the prefix `optimizer_`, e.g., adamax can be called via `optimizer_adamax`

Level	Name	Description	Reference
1	sgd	SGD optimizer with support for momentum, learning rate decay, and Nesterov momentum	Ruder (2017)
2	rmsprop	RMSProp optimizer	Ruder (2017)
3	adagrad	Adagrad optimizer	Duchi et al. (2011)
4	adadelta	Adadelta optimizer	Zeiler (2012)
5	adam	Adam optimizer	Kingma and Ba (2014)
6	adamax	Adamax optimizer	Kingma and Ba (2014)
7	nadam	Nesterov Adam optimizer	Sutskever et al. (2013)

To enable compatibility with the ranges of the learning rates of the other optimizers, the learning rate of the optimizer `adadelta` is internally mapped to `1-learning_rate`. That is, a learning rate of 0 will be mapped to 1 (which is `adadelta`'s default learning rate). The learning rate of `adagrad` and `sgd` is internally mapped to `10 * learning_rate`. That is, a learning rate of 0.001 will be mapped to 0.01 (which is `adagrad`'s and `sgd`'s default). The learning rate learning_rate of `adamax` and `nadam` is internally mapped to `2 * learning_rate`. That is, a learning rate of 0.001 will be mapped to 0.002 (which is `adamax`'s and `nadam`'s default.)

The hyperparameter x_{11}, which encodes the `optimizer` is implemented as a factor. Factor levels, which represent the available optimizers are listed in Table 10.5.

A discussion of the DNN hyperparameters, λ, recommendations for their settings and further information are presented in Sect. 3.8. The R function `getModelConf` provides information about hyperparameter names, ranges, and types.

10.3.3 The Neural Network

Background: Network Implementation in SPOTMisc

The `SPOTMisc` function `getModelConf` selects a pre-specified, but not pretrained, DL network $\mathcal{A}$. This network is called via `funKerasGeneric`, which is the interface to `spot`. `funKerasGeneric` uses a network, that is implemented as follows:

To build the DNN in `keras`, the function `layer_dense_features` that processes the feature columns specification is used (Fig. 10.2). It receives the data set `specGeneric_prep` as input and returns an array off all dense features:

```
layer <-
  layer_dense_features(
    feature_columns = dense_features(specList$specGeneric_prep)
  )
```

The iterator can be called to take a look at the (scaled) output:

```
specList$train_ds_generic %>%
  reticulate::as_iterator() %>%
  reticulate::iter_next() %>%
  layer()
```

The NN model can be compiled after the `loss` function $\mathcal{L}$, which determines how good the DNN prediction is (based on the $(X, Y)^{(\text{val})}$), the `optimizer`, i.e., O_{inner}, i.e., the update mechanism of $\mathcal{A}$, which adjusts the weights using backpropagation, and the `metrics`. *metrics* The metrics monitor the progress during training and testing and are specified using the `compile` function from `keras`.

! Attention: Hyperparameter Values

To improve the readability of the code, evaluated ("forced" values) of the hyperparameters λ are shown in the code snippets below instead of the arguments that are passed from the tuner `spot` to the function `funKerasGeneric`.

```
units1 <- 2
model <- keras_model_sequential() %>%
  layer_dense_features(dense_features(specList$specGeneric_prep)) %>%
  layer_dense(units = units1, activation = "relu") %>%
  layer_dense(units = 1, activation = "sigmoid")
model %>% compile(
  loss = loss_binary_crossentropy,
  optimizer = "adam",
  metrics = "binary_accuracy"
)
```

The DNN training can be started as follows (using `keras`' `fit` function). Train the model on the CPU using the setting `tf$device("/cpu:0")` on the validation data set:

```
with(tf$device("/cpu:0"), {
  historyD <-
    model %>%
    fit(dataset_use_spec(specList$train_ds_generic,
      spec = specList$specGeneric_prep
    ),
    epochs = 25,
```

Fig. 10.2 Simple DNN
based on the code in this
section

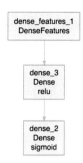

```
    validation_data =
      dataset_use_spec(
        specList$val_ds_generic,
        specList$specGeneric_prep
      ),
    verbose = 0
    )
})
```

The predictions from the DNN model are shown in the following code snippet.
The tensor values are the output from the final DNN layer after the sigmoid function
was applied. Values are from the interval [0, 1] and represent probabilities: values
smaller than 0.5 are interpreted as predictions "age < 40", otherwise "age ≥ 40".

```
specList$test_ds_generic %>%
  reticulate::as_iterator() %>%
  reticulate::iter_next() %>%
  model()
  ## tf.Tensor(
  ## [[0.31883082]
  ##  [0.47055224]
  ##  [0.99928933]
  ##  [0.9962864 ]
  ##  [0.27977774]
  ##  [0.34997565]
  ##  [0.7686823 ]
  ##  [0.99928933]
  ##  [0.32100695]
  ##  [0.99928933]
  ##  [0.16852783]
  ##  [0.33614054]
  ##  [0.36855838]
  ##  [0.4346528 ]
  ##  [0.6968227 ]
  ##  [0.41458437]
```

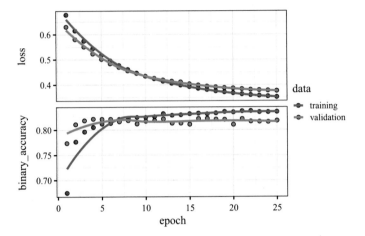

Fig. 10.3 DNN training. History of the inner optimization loop

```
##    [0.9992539 ]
##    [0.01920704]
##    [0.34810022]
##    [0.6455758 ]
##    [0.78468007]
##    [0.9993542 ]
##    [0.9469396 ]
##    [0.00989294]
##    [0.00521746]
##    [0.01071302]
##    [0.8888161 ]
##    [0.78542197]
##    [0.9993542 ]
##    [0.6045456 ]
##    [0.9993542 ]
##    [0.9992539 ]], shape=(32, 1), dtype=float32)
```

Figure 10.3 shows the quantities that are being displayed during training:

(i) the *loss* of the network over the training and validation data, $\psi^{(\text{train})}$ and $\psi^{(\text{val})}$, respectively, and

(ii) the *accuracy* of the network over the training and validation data, $f_{\text{acc}}^{(\text{train})}$ and $f_{\text{acc}}^{(\text{val})}$, respectively.

This figure illustrates that an accuracy greater than 80% on the training data, $(X, Y)^{(\text{train})}$, can be reached quickly.

Figure 10.3 can indicate (even if this is only a short fit procedure) whether the modeling is affected by overfitting or not. If this situation occurs, it might be useful to implement dropout layers or use other methods to prevent overfitting.

The effects of HPT and the tunability of $\mathcal{A}$ will be described in the following sections. Finally, using `keras`' `evaluate` function, the DNN model performance can be checked on $X^{(\text{test})}$.

```
model %>%
  evaluate(specList$test_ds_generic %>%
    dataset_use_spec(specList$specGeneric_prep), verbose = 0)

##              loss binary_accuracy
##         0.3550636       0.8068387
```

The relationship between $\psi^{(\text{train})}$, $\psi^{(\text{val})}$, and $\psi^{(\text{test})}$ as well as between $f_{\text{acc}}^{(\text{train})}$, $f_{\text{acc}}^{(\text{val})}$, and $f_{\text{acc}}^{(\text{test})}$ can be analyzed with Sequential Parameter Optimization Toolbox (SPOT), because it computes and reports these values.

10.4 `funKerasGeneric`: The Objective Function

The hyperparameter tuner, e.g., `spot`, performs model selection during the tuning run: training data $X^{(\text{train})}$ is used for fitting (training) the models, e.g., the weights of the DNNs. Each trained model $\mathcal{A}_{\lambda_i}\left(X^{(\text{train})}\right)$ will be evaluated on the validation data $X^{(\text{val})}$, i.e., the loss is calculated as shown in Eq. (2.9). Based on $(\lambda_i, \psi_i^{(\text{val})})$, at each iteration of the outer optimization loop a surrogate model $\mathcal{S}(t)$ is fitted, e.g., a Bayesian Optimization (BO) (Kriging) model using `spot`'s `buildKriging` function.

For each hyperparameter configuration λ_i, the objective function `funKerasGeneric` reports information about the related DNN models $\mathcal{A}_{\lambda_i}$

1. training loss, $\psi^{(\text{train})}$,
2. training accuracy, $f_{\text{acc}}^{(\text{train})}$,
3. validation (testing) loss, $\psi^{(\text{val})}$, and
4. validation (testing) accuracy, $f_{\text{acc}}^{(\text{val})}$.

10.5 `spot`: Experimental Setup for the Hyperparameter Tuner

The SPOT package for R, which was introduced in Sect. 4.5, will be used for the DL hyperparameter tuning (Bartz-Beielstein et al. 2021). The budget is set to twelve hours, i.e., the run time of DL tuning is larger than the run time of the ML tuning. The budget for the `spot` runs was set to this value, because of the complexity of the hyperparameter search space Λ and the relatively long run time of the DNN.

SPOT provides several options for adjusting the HPT parameters, e.g., type of the Surrogate Model Based Optimization (SMBO) model, $\mathcal{S}$, and optimizer, $\mathcal{O}$, as well as the size of the initial design, n_{init}. These parameters can be passed via the `spotControl` function to `spot`. For example, instead of the default surrogate $\mathcal{S}$, which is BO (implemented as `buildKriging`), a Random Forest (RF), (implemented as `buildRanger`) can be chosen.

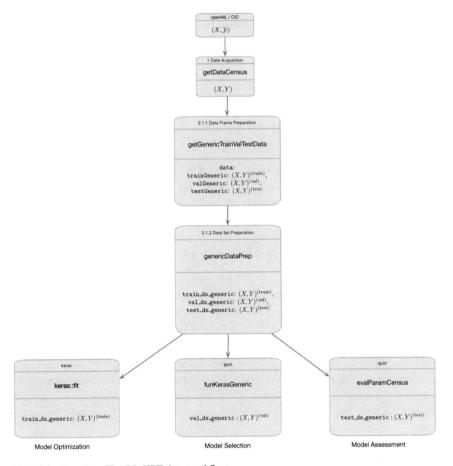

Fig. 10.4 Overview. The DL HPT data workflow

The general DL HPT data workflow is as follows: first the training data, $(X, Y)^{(train)}$ are fed to the DNN. The DNN will then learn to associate images and labels. Based on the `keras` parameter `validation_split`, the training data will be partitioned into a (smaller) training data set, $X^{(train)}$, and a validation data set, $(X, Y)^{(valtrain)}$. The trained DNN produces predictions for validations based on $(X, Y)^{(val)}$ data. The DL HPT data workflow is shown in Fig. 10.4.

Similar to the process described in Sect. 8.1 for ML, the hyperparameter tuning for DL can be started as follows:

startCensusRun(model = "dl")

The `startCensusRun` function performs the following steps:

1. Providing the CID data set, $((X, Y)_{CID}$, see Sect. 8.2.1.
2. Generating the random sample $(X, Y) \subseteq ((X, Y)_{CID}$ of size `nobs`.

Table 10.6 SPOT parameters used for deep learning hyperparameter tuning. The `control` list contains internally further lists, see Table 10.7

Parameter	Value	Description
x	x0	Starting point, hyperparameter vector λ, see Tables 10.4 and 10.5
fun	funKerasGeneric	Objective function, O_{outer}
lower	cfg$lower	Lower bounds for x aka λ
upper	cfg$upper	Upper bounds for x aka λ
control	List	
kerasConf	kerasConf	Argument used by the objective function funKerasGeneric
specList	specList	Argument used by the objective function funKerasGeneric

Table 10.7 SPOT list parameters used for deep learning hyperparameter tuning

List	Parameter	Value
Control	Types	cfg$type
	Verbosity	Verbosity
	Time	List (maxTime = timebudget/60)
	Plots	Plots
	Progress	TRUE
	Model	spotModel
	Optimizer	spotOptim
	Noise	Noise
	OCBA	OCBA
	OCBABudget	OCBABudget
	seedFun	NA
	seedSPOT	tuner.seed
designControl	Replicates	Rinit
	Size	initSizeFactor * length(cfg$lower)
modelControl	Target	krigingTarget
	useLambda	krigingUseLambda
	Reinterpolate	krigingReinterpolate
optimizerControl	funEvals	multFun * length(cfg$lower))
yImputation	handleNAsMethod	handleNAsMethod
	imputeCriteriaFuns	imputeCriteriaFuns
	penaltyImputation	3

3. Defining an experimental design, including performance measures.
4. Configuration of the hyperparameter tuner, $\mathcal{T}$.
5. Configuration of the DL model, $\mathcal{A}$.
6. Performing the experiments.

Furthermore, it can be decided whether to use the default hyperparameter setting, λ_0, as a starting point or not. Using the parameter specifications from Tables 10.6 and 10.7, we are ready to perform the HPT run: `spot` can be started.

10.6   Tunability

Regarding tunability as defined in Definition 2.26, we are facing a special situation in this chapter, because there is no generally accepted "default" hyperparameter configuration, λ_0, for DNNs. This problem is not as obvious in ML, because the corresponding methods have a long history, i.e., there are publications for most of the shallow methods that can give hints how to select adequate λ values. This information is collected and summarized in Chap. 3. The "default" hyperparameter setting of the DNNs analyzed in this chapter is based on our own experiences, combined with recommendations in the literature. Chollet and Allaire (2018) may be considered as a reference in this field.[7]

The `result` list from the `spot` run can be loaded. It contains the 14 values shown in Table 4.6, e.g., names of the tuned hyperparameters that were introduced in Table 10.4:

```
result$control$parNames
##  [1] "dropout"       "dropoutfact"  "units"        "unitsfact"
##  [5] "learning_rate" "epochs"       "beta_1"       "beta_2"
##  [9] "layers"        "epsilon"      "optimizer"
```

The HPT inner optimization loop is shown in Fig. 10.5. The DNN uses the tuned hyperparameters, $\lambda^\star$ from Table 10.8. The model training supports the result found by the tuner `spot` that the number of training epochs should be 32. The reader may compare the inner optimization loop with default and with tuned hyperparameters in Figs. 10.3 and 10.5.

The tuned DNN model has the following structure:

```
## $model
## Model: "sequential_1"
##
## _____
## Layer (type)                     Output Shape              Param #
## ========================================================================
## dense_features_2 (DenseFeatures)  multiple                 0
```

---

[7] An updated version of Chollet and Allaire (2018) is under preparation while we are writing this text. Check the authors' web-page for more information: https://www.manning.com/books/deep-learning-with-r.

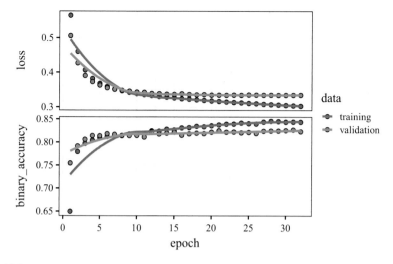

Fig. 10.5 Training DL (inner optimization loop) using the tuned hyperparameter setting $\lambda^\star$

Table 10.8 DNN configurations. "lr" denotes "learning_rate". The overall mean of the loss, $\overline{y}$ is 0.3691, its standard deviation is 0.1152, whereas the mean of the best HPT configuration, $\lambda^\star$, found by OCBA, is 0.3346 with s.d. 0.0343

dropout	dropoutfact	units	unitsfact	lr	epochs	beta_1	beta_2	layers	epsilon	optimizer	Loss
0	0	5	0.5	0.001	4	0.9	0.999	1	0	5	0.346
0.038	0.793	5	0.742	0.002	5	0.913	0.994	1	0	4	0.335

```
## dense_2 (Dense)                    multiple                        8864
## dense_3 (Dense)                    multiple                        33
## ===========================================================================
## Total params: 8,897
## Trainable params: 8,897
## Non-trainable params: 0
##
##
## _____
##
## $history
##
## Final epoch (plot to see history):
##                  loss: 0.2983
##       binary_accuracy: 0.8508
##              val_loss: 0.3343
## val_binary_accuracy: 0.8132
```

10.6.1 Progress

After loading the results from the experiments, the hyperparameter tuning progress can be visually analyzed. First of all, the `result` list information will be used to

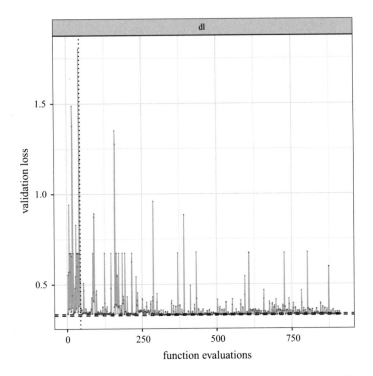

Fig. 10.6 Progress plot. In contrast to the progress plots used for the ML methods, this plot shows the BCE loss and not the MMCE against the number of iterations (function evaluations of the tuner)

visualize the *route to the solution*: in Fig. 10.6, loss function values, $\psi^{(\text{val})}$, are plotted against the number of iterations, t. Each point represents one evaluation of an DNN model $\mathcal{A}_\lambda(t)$ at time step (spot iteration) t.

The initial design, which includes the default hyperparameter setting, λ_0, results in a loss value of $\psi_{\text{init}}^{(\text{val})} = 0.3371$. The best value, that was found during the tuning, is $y_{\text{val}}^{(*)} = 0.3285$. These values have to be taken with caution, because they represent onyl one evaluation of $\mathcal{A}_\lambda$. Based on OCBA, which takes the noise in the model evaluation via the function funKerasGeneric into consideration, the best function value is $y_{\text{val}}^{(\text{OCBA}^*)} = 0.3346$.

After 12 h, 914 dl models were evaluated. Comparing the worst configuration that was observed during the HPT with the best, a 81.773% reduction in the BCE loss was obtained. After the initial phase, which includes 44 evaluations, the smallest BCE reads 0.3370858. The dotted red line in Fig. 8.6 illustrates this result. The final best value reads 0.3285304, i.e., a reduction of the BCE of 2.5381%. These values, in combination with results shown in the progress plot (Fig. 8.6) indicate that a relatively short HPT run is able to improve the quality of the DNN model. It also indicates, that increased run times do not result in a significant improvement of the BCE. The full

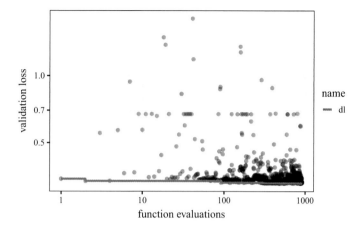

Fig. 10.7 Log-log plot

comparison of the DL and ML algorithm performances with default, λ_0, and tuned, $\lambda^\star$, hyperparameters is shown in Sect. 10.9.

> **! Attention**
>
> These results do not replace a sound statistical comparison, they are only indicators, not final conclusions.

The corresponding code is presented in the Appendix. The related hyperparameters values are shown in Table 10.8.

There is a large variance in the loss as can be seen in Figs. 10.6 and 10.7. The latter of these two plots visualizes the same data as the former, but uses log-log axes instead.

10.6.2   *evalParamCensus: Comparing Default and Tuned Parameters on Test Data*

The function evalParamCensus evaluates ML and DL hyperparameter configurations on the CID data set. It compiles a data frame, which includes performance scores from several hyperparameter configurations and can also process results from default settings. This data frame can be used for a comparison of default and tuned hyperparameters, λ_0 and $\lambda^\star$, respectively. A violin plot of this comparison is shown in Fig. 10.8. It is based on 30 evaluations of λ_0 and $\lambda^\star$ and shows—in contrast to the values in the DNN progress plots—the Mean Mis-Classification Error (MMCE). The MMCE was chosen to enable a comparison of the DL results with the ML results

Fig. 10.8 Comparison of DL algorithms with default (D) and tuned (T) hyperparameters. Mean misclassification error (MMCE) for both configurations. Vertical lines mark quantiles (0.25, 0.5, 0.75) of the corresponding distribution. Numerical values are shown in Table 10.8

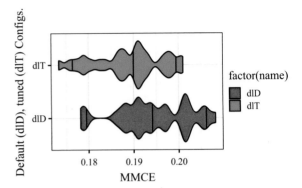

shown in this book. Identical evaluations were done in Chaps. 8, 9, and 12. A global comparison of the six ML and the DL methods from this book will be shown in Sect. 10.9.

10.7 Analysing the Deep Learning Tuning Process

The values that are used for the analysis in this section are biased because they are not using an experimental design (space filling or factorial). Instead, they are using the data from the `spot` tuning process, i.e., they are biased by the search strategy (Expected Improvement (EI)) on the surrogate S.

Identical to the analysis of the ML methods, a simple regression tree as shown in Fig. 10.9 can be used for analysing effects and interactions between hyperparameters λ.

The regression tree supports the observations, that units and epochs have the largest effect on the validation loss. The importance of the parameters from the random forest analysis are shown in Table 10.9.

To perform a sensitivity analysis, parallel and sensitivity plots can be used.

The parallel plot (Fig. 10.10) indicates that the hyperparameter `units` should be set to a value of 32 (the transformed values range from 1 to 32), the `epochs`, i.e. x_6, should be set to a value of 32 (the transformed values range from 8 to 128), the `layers`, i.e. x_9, should be set to a value of 1 (the transformed values range from 1 to 4), and the `optimizer`, i.e. x_{11}, should be set to a value of 4 (the transformed values range from 1 to 7).

Looking at Fig. 10.11, the following observations can be made: Similar to the results from the parallel plot (Fig. 10.10), the sensitivity plot shows that the `epochs`, i.e. x_6, and the `optimizer`, i.e. x_{11}, have the largest effect: the former leads to poor results for larger values, whereas the latter produces poor results for relatively small values. This indicates that the number of training epochs should not be too large

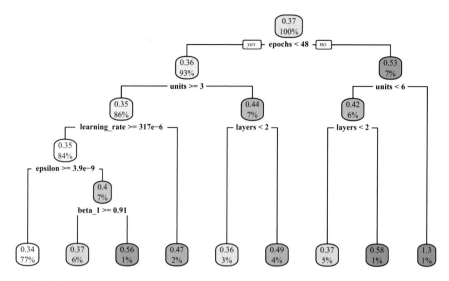

Fig. 10.9   Regression tree. Deep learning model. Transformed hyperparameter values are shown

Table 10.9   Variable importance of the DL model hyperparameters

λ_i	units	epochs	beta_2	layers	lr	beta_1	eps	opt.	dropoutfact	dropout	unitsfact
Var. imp.	6.04	1.69	1.36	0.63	0.47	0.46	0.42	0.33	0.22	0.19	0.10

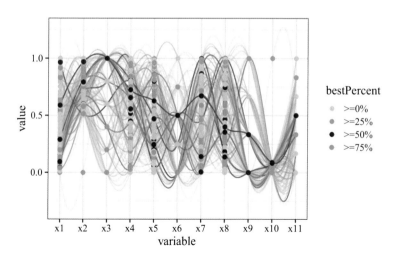

Fig. 10.10   Best configurations in green

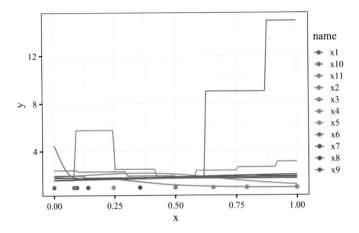

Fig. 10.11 Sensitivity plot (best)

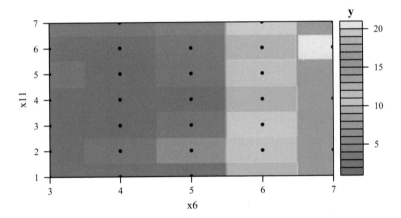

Fig. 10.12 Surface plot: epochs x_6 plotted against optimizer x_{11}. This plot indicates that longer training (larger epochs values) worsen the performance and that the optimizer `adadelta` performs well. Note: Plateaus are caused by discrete and factor variables

(probably to prevent overfitting, see Fig. 10.5) and that the optimizers `adadelta` or `adam` are recommended (Fig. 10.12).

Finally, a simple linear regression model can be fitted to the data. Based on the data from SPOT's `res` list, this can be done as follows:

```
##
## Call:
## lm(formula = y ~ ., data = df)
##
## Residuals:
##       Min       1Q    Median       3Q       Max
```

```
## -0.19062 -0.04055 -0.00477 -0.00044  1.16255
##
## Coefficients:
##                 Estimate Std. Error t value Pr(>|t|)
## (Intercept)    2.291e+00  1.494e+00   1.533 0.125532
## dropout        9.491e-02  3.804e-02   2.495 0.012776 *
## dropoutfact    3.807e-02  3.167e-02   1.202 0.229606
## units          9.670e-03  3.544e-03   2.729 0.006484 **
## unitsfact     -5.514e-02  2.225e-02  -2.478 0.013396 *
## learning_rate  8.281e+00  1.509e+00   5.488 5.29e-08 ***
## epochs         4.832e-02  4.628e-03  10.442  < 2e-16 ***
## beta_1         3.456e-01  1.705e-01   2.028 0.042888 *
## beta_2        -2.589e+00  1.486e+00  -1.743 0.081739 .
## layers         2.360e-02  4.573e-03   5.161 3.03e-07 ***
## epsilon       -1.522e+06  7.284e+05  -2.089 0.036961 *
## optimizer2     4.672e-02  1.783e-02   2.620 0.008933 **
## optimizer3     2.791e-02  1.575e-02   1.772 0.076659 .
## optimizer4    -9.552e-03  1.343e-02  -0.711 0.477196
## optimizer5     1.282e-01  2.094e-02   6.121 1.39e-09 ***
## optimizer6     6.941e-02  1.572e-02   4.415 1.13e-05 ***
## optimizer7     1.172e-01  3.020e-02   3.880 0.000112 ***
## ---
## Signif. codes:  0 '***' 0.001 '**' 0.01 '*' 0.05 '.' 0.1 ' ' 1
##
## Residual standard error: 0.09997 on 897 degrees of freedom
## Multiple R-squared:  0.2595, Adjusted R-squared:  0.2463
## F-statistic: 19.65 on 16 and 897 DF,  p-value: < 2.2e-16
```

Although this linear model requires a detailed investigation (a misspecification analysis is recommended, see, e.g., Spanos 1999), it can be used in combination with other Exploratory Data Analysis (EDA) tools and visualizations from this section to discover unexpected and/or interesting effects. It should not be used alone for a final decision. Despite of a relatively low adjusted R^2 value, the regression output shows—in correspondence with previous observations—that increasing the number of epochs worsens the model performance.

10.8 Severity: Validating the Results

Considering the results of the experimental runs the difference is $\bar{x} = 0.0054$. Since this value is positive, for the moment, let us assume that the tuned solution is superior. The corresponding standard deviation is $s_d = 0.0056$. Based on Eq. 5.14, and with $\alpha = 0.05$, $\beta = 0.2$, and $\Delta = 0.006$.

Next, we will identify the required number of runs for the full experiment using the getSampleSize function. For a relevant difference of 0.006 approximately 11 completing runs per algorithm are required. Hence, we can directly proceed to evaluate the severity and analyse the performance improvement achieved through tuning the parameters of the DL model.

Table 10.10 Case Study III: Result Analysis

p-value	Decision	Power	Cohen's d	Hedge's g	Severity
0	H0 rejected	0.9999849	0.695314	0.686284	$\Delta \leq 0.0045$ are well supported

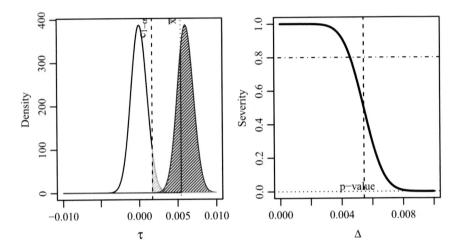

Fig. 10.13 Tuning DL. Severity of rejecting H0 (red), power (blue), and error (gray). Left: the observed mean $\bar{x} = 0.0054$ is larger than the cut-off point $c_{1-\alpha} = 0.0017$ Right: The claim that the true difference is as large or larger than 0.006 is not supported by severity. But, any difference smaller than 0.0045 is supported by severity

Result summaries are presented in Table 10.10. The decision based on p-value is to reject the null hypothesis, i.e, the claim that the tuned parameter setup provides a significant performance improvement in terms of MMCE is supported. The effect size suggests that the difference is of medium magnitude. For the chosen $\Delta = 0.006$, the severity value is at 0.29 and thus it does not support the decision of rejecting the H_0. The severity plot is shown in Fig. 10.13. Severity shows that only performance differences smaller than 0.0045 are well supported.

10.9 Summary and Discussion

A HPT approach based on SMBO was introduced and exemplified in this chapter. It uses functions from the packages `keras`, `SPOT` and `SPOTMisc` from the statistical programming environment R, hence providing a HPT environment that is fully accessible from R. Although HPT can be performed with R functions, an underlying PYTHON environment has to be installed. This installation is explained in the Appendix.

The first three case studies in this book are concluded with a global comparison of the seven methods, i.e., six ML methods and one DL method. The main goal of these studies was to analyze whether a relatively short HPT run, which is performed on a notebook or desktop computer without High Performance Computing (HPC) hardware, can improve the performance. Or, stated differently:

Is it worth doing a short HPT run before doing a longer study?

To illustrate the performance gain (tunability), a final comparison of the seven methods will be presented. The number of repeats will be determined first:

An approximate formula for sample size determination will be used. The reader is referred to Sect. 5.6.5 and to Senn (2021) for details. A sample size of 30 experiments was chosen, i.e., altogether 210 runs were performed.

The list of results from the rfunctionspot HPT run stores relevant information about the configuration and the experimental results.

Violin plots (Fig. 10.14) can be used. These observations are based on data collected from default and tuned parameter settings. Although the absolute best value was found by Extreme Gradient Boosting (XGBoost), Support Vector Machine (SVM) should be considered as well, because the performance is similar while the variance is much lower. This study briefly explained how HPT can be used as a datascope for the optimization of DNN hyperparameters. The results from this brief study scratch on the surface of the HPT set of tools. Especially for DL, SPOT allows recommendations for improvement, it provides tools for comparisons using different losses and measures on different data sets, e.g., $\psi^{(train)}$, $\psi^{(val)}$, and $\psi^{(test)}$.

While discussing the hyperparameter tuning results, HPT does not search for the final, best solution only. For sure, the hyperparameter practitioner is interested in the best solution. But even from this *greedy* point of view, considering the *route to the solution* is also of great importance, because analyzing this route enables *learning* and can be much more efficient in the long run compared to a greedy strategy.

Example: Route to the solution

Consider a classification task that has to be performed several times in a different context with similar data. Instead of blindly (automatically) running the Hyperparameter Optimization (HPO) procedure individually for each classification task (which might also require a significant amount of time and resources, even when it is performed automatically) a few HPT procedures are performed. Insights gained from HPT might help to avoid ill specified parameter ranges, too short run times, and further pitfalls.

In addition to an effective and efficient way to determine the optimal hyperparameters, SPOT provides means for understanding algorithms' performance (we will use

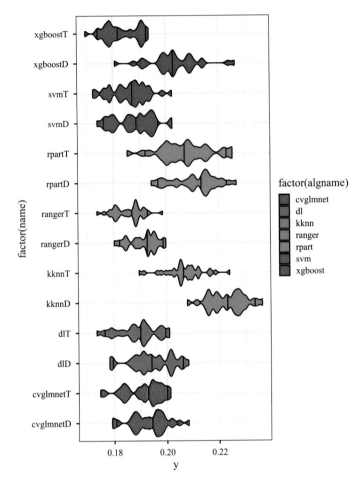

Fig. 10.14 Comparison of ML algorithms with default (D) and tuned (T) hyperparameters. Classification error (MMCE). Note: because there is no "default" hyperparameter setting for the deep learning models used in this study, we have chosen a setting based on our experience and recommendations from the literature, see the discussion in Sect. 10.6

datascopes similar to microscopes in biology and telescopes in astronomy). Considering the research goals stated in Sect. 4.1, the HPT approach presented in this study provides many tools and solutions.

To conclude this chapter, in addition to the research goals (R-1) to (R-8) from Sect. 4.1, important goals that are specific for HPT in DNN are presented.

The selection of an adequate performance measure is relevant. Kedziora et al. (2020) claimed that "research strands into ML performance evaluation remain arguably disorganized, [...]. Typical ML benchmarks focus on minimizing both loss functions and processing times, which do not necessarily encapsulate the entirety of human requirement." Furthermore, a sound test problem specification is necessary, i.e., train, validation, and test sets should be clearly specified. Importantly, the initialization (this is similar to the specification of starting points in optimization) procedures should be made transparent. Because DL methods require a large amount of computational resources, the usage of surrogate benchmarks should be considered (this is similar to the use of Computational Fluid Dynamics (CFD) simulations in optimization). Most of the ML and DL methods are noisy. Therefore, repeats should be considered. The power of the test, severity, and related tools which were introduced in Chap. 5 can give hints for choosing adequate values, i.e., how many runs are feasible or necessary. The determination of meaningful differences—with respect to the specification of the loss function or the accuracy—based on tools like severity are of great relevance for the practical application. Remember: scientific relevance is not identical to statistical significance. Furthermore, floor and ceiling effects should be avoided, i.e., the comparison should not be based on too hard (or too easy) problems. We strongly recommend a comparison to baseline (e.g., default settings or Random Search (RS)).

The model $\mathcal{A}$ must be clearly specified, i.e., the initialization, pre-training (starting points in optimization) should be explained. The hyperparameter (ranges, types) should be clearly specified. If there are any additional (untunable) parameters, then they should be explained. How is reproducibility ensured (and by whom)? Last but not least: open source code and open data should be provided.

The final conclusion from the three case studies (Chaps. 8–10) can be formulated as follows:

> HPT provides tools for comparing, analyzing, and selecting an adequate ML or DL method for unknown real-world problems. It requires only moderate computational resources (notebooks or desktop computers) and limited time. Practitioners can start HPT runs at the end of their work day and will find the results ready on their desk the next morning.

10.10   Program Code

Program Code

```
runNr <- "000"
batch_size <- 16
prop <- 2 / 3
dfGeneric <- getDataCensus(target = target, nobs = 1000)
# dfGeneric <- MASS::Boston
# names(dfGeneric)[names(dfGeneric) == "medv"] <- "target"
```

```
data <- getGenericTrainValTestData(dfGeneric = dfGeneric, prop = prop)
specList <- genericDataPrep(data = data, batch_size = batch_size)
## model configuration:
model <- "dl"
cfg <- getModelConf(list(model = model))
x <- matrix(cfg$default, nrow = 1)
#'
kerasConf <- getKerasConf()
kerasConf$nClasses <- 1
kerasConf$activation <- NULL
kerasConf$verbose <- 0
kerasConf$loss <- "mse"
kerasConf$metrics <- "mae"
##  Only some variables are tuned
# kerasConf$active <-  c("layers", "units", "epochs")
### First example: simple function call:
message("objectiveFunctionEvaluation(): x before transformX().")
print(x)
if (length(cfg$transformations) > 0) {
  x <- transformX(xNat = x, fn = cfg$transformations)
}
message("objectiveFunctionEvaluation(): x after transformX().")
print(x)
funKerasGeneric(x, kerasConf = kerasConf, specList = specList)
#'
### Second example: evaluation of several (three) hyperparameter settings:
xxx <- rbind(x, x, x)
funKerasGeneric(xxx, kerasConf = kerasConf, specList)
#'
### Third example: spot call
kerasConf$verbose <- 0
result <-
  spot(
    x = NULL,
    fun = funKerasGeneric,
    lower = cfg$lower,
    upper = cfg$upper,
    control = list(
      funEvals = 25,
      # time = list(maxTime = 5),
      noise = TRUE,
      types = cfg$type,
      plots = TRUE,
      progress = TRUE,
      seedFun = 1,
      seedSPOT = 1,
      replicates = 2,
      OCBA = TRUE,
      OCBABudget = 2,
      parNames = cfg$tunepars,
      designControl = list(
        replicates = 2,
        size = 1 * length(cfg$lower)
      ),
      yImputation = list(
        handleNAsMethod = handleNAsMean,
        imputeCriteriaFuns = list(is.infinite, is.na, is.nan),
        penaltyImputation = 3
      ),
```

```r
    modelControl = list(
      target = "ei",
      useLambda = TRUE,
      reinterpolate = FALSE
    ),
    transformFun = cfg$transformations
  ),
  kerasConf = kerasConf,
  specList = specList
)
x <- result$xbest
message("objectiveFunctionEvaluation(): x before transformX().")
print(x)
if (length(cfg$transformations) > 0) {
  x <- transformX(xNat = x, fn = cfg$transformations)
}
message("objectiveFunctionEvaluation(): x after transformX().")
print(x)
df <- data.frame(x)
names(df) <- cfg$tunepars
print(df)
save(result, file = paste0(model, runNr, ".RData"))


dfRun <- prepareProgressPlot(model, runNr, directory = ".")
ggplotProgress(dfRun)


library("rpart")
library("rpart.plot")
library("SPOT")
x <- result$x
# cfg <- getModelConf(model="dl")
transformFun <- cfg$transformations
message("predDlCensus(): x before transformX().")
print(x)
if (length(cfg$transformations) > 0) {
  x <- transformX(xNat = x, fn = cfg$transformations)
}
message("predDlCensus(): x after transformX().")
print(xt)
fitTree <- buildTreeModel(
  x = xt,
  y = result$y,
  control = list(xnames = result$control$parNames)
)
rpart.plot(fitTree$fit)


kerasConf$returnObject <- "pred"
x <- result$xbest
if (length(cfg$transformations) > 0) {
  x <- transformX(xNat = x, fn = cfg$transformations)
}
x
evalKerasGeneric(
```

```r
  x = x,
  kerasConf = kerasConf,
  specList = specList
)


library("SPOT")
library("SPOTMisc")
runNr <- "OCBA"
batch_size <- 32
prop <- 2 / 3
target <- "age"
dfGeneric <- getDataCensus(target = target, nobs = 1e4)
# dfGeneric <- MASS::Boston
# names(dfGeneric)[names(dfGeneric) == "medv"] <- "target"
data <- getGenericTrainValTestData(dfGeneric = dfGeneric, prop = prop)
specList <- genericDataPrep(data = data, batch_size = batch_size)
## model configuration:
model <- "dl"
cfg <- getModelConf(list(model = model))
# x <- matrix(cfg$default, nrow=1)
x <- result$xBestOcba
#'
kerasConf <- getKerasConf()
# kerasConf$nClasses <- 1
# kerasConf$activation <- NULL
kerasConf$verbose <- 2
# kerasConf$loss <- "mse"
# kerasConf$metrics <- "mae"
##  Only some variables are tuned
# kerasConf$active <-  c("layers", "units", "epochs")
### First example: simple function call:
message("objectiveFunctionEvaluation(): x before transformX().")
print(x)
if (length(cfg$transformations) > 0) {
  x <- transformX(xNat = x, fn = cfg$transformations)
}
message("objectiveFunctionEvaluation(): x after transformX().")
print(x)
evalKerasGeneric(
  x = x,
  kerasConf = kerasConf,
  specList = specList
)
```

Chapter 11
Case Study IV: Tuned Reinforcement Learning (in PYTHON)

Martin Zaefferer and Sowmya Chandrasekaran

Abstract Similar to the example in Chap. 10, which considered tuning a Deep Neural Network (DNN), this chapter also deals with neural networks, but focuses on a different type of learning task: reinforcement learning. This increases the complexity, since any evaluation of the learning algorithm also involves the simulation of the respective environment. The learning algorithm is not just tuned with a static data set, but rather with dynamic feedback from the environment, in which an agent operates. The agent is controlled via the DNN. Also, the parameters of the reinforcement learning algorithm have to be considered in addition to the network parameters. Based on a simple example from the Keras documentation, we tune a DNN used for reinforcement learning of the inverse pendulum environment toy example. As a bonus, this chapter shows how the demonstrated tuning tools can be used to interface with and tune a learning algorithm that is implemented in PYTHON.

Supplementary Information The online version contains supplementary material available at https://doi.org/10.1007/978-981-19-5170-1_11.

M. Zaefferer (✉)
Bartz & Bartz GmbH and with Institute for Data Science, Engineering, and Analytics, TH Köln, Gummersbach, Germany

Duale Hochschule Baden-Württemberg Ravensburg, Ravensburg, Germany
e-mail: zaefferer@dhbw-ravensburg.de

S. Chandrasekaran
Institute for Data Science, Engineering, and Analytics, TH Köln, Steinmüllerallee 1, 51643 Gummersbach, Germany
e-mail: sowmya.chandrasekaran@th-koeln.de

11.1 Introduction

In this chapter, we will demonstrate how a reinforcement learning algorithm can be tuned. In reinforcement learning, we consider a dynamic learning process, rather than a process with fixed, static data sets like in typical classification tasks.

The learning task considers an *agent*, which operates in an *environment*. In each timestep, the *agent* decides to take a certain *action*. This *action* is fed to the *environment*, and causes a change from a previous *state* to a new *state*. The *environment* also determines a *reward* for the respective *action*. After that, the *agent* will decide the next *action* to take, based on received *reward* and the new *state*. The learning goal is to find an agent that accumulates as much reward as possible.

To simplify things for a second, let us consider an example: A mobile robot (*agent*) is placed in a room (*environment*). The *state* of the agent is the position of the robot. The *reward* may be based on the distance traveled toward a target position. Different movements of the robot are the respective *actions*.

In this case, our neural network can be used to map from the current state to a new action. Thus, it presents a controller for our robot agent. The weights of this neural network have to be learned in some way, taking into account the received rewards. Compared to Chap. 10, this leads to a somewhat different scenario: Data is usually gathered in a dynamic process, rather than being available from the start.[1] In fact, initially, we may not have any data. We acquire data during the learning process, by observing states/actions/rewards in the environment.

11.2 Materials and Methods

11.2.1 Software

We largely rely on the same software as in the previous chapters. That is, we use the same tuning tools. As in Chap. 10, we use Keras and TensorFlow to implement the neural networks. However, we will perform the complete learning task within PYTHON, using the R package `reticulate` to explicitly interface between the R-based tuner and the PYTHON-based learning task (rather than implicitly via R's `keras` package).

On the one hand, this will demonstrate how to interface with different programming languages (i.e., if your model is not trained in R). On the other hand, this is a necessary step, because the respective environment is only available in PYTHON (i.e., the toy problem).

For the sake of readability, the complete code will not be printed within the main text, but is available as supplementary material.

---

[1] Although it has to be noted that somewhat similar dynamics may occur, e.g., when learning a classification model with streaming data.

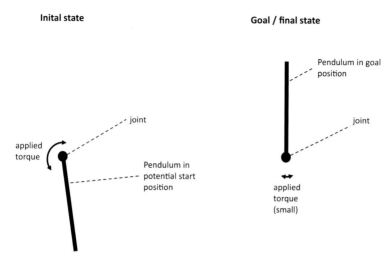

Fig. 11.1 Initial and goal state of the 2D inverted pendulum problem

11.2.2 Task Environment: Inverted Pendulum

The example we investigate is based on a Keras tutorial by Singh (2020). This relies on a toy problem that is often used to test or demonstrate reinforcement learning algorithms: the frictionless inverted pendulum. More specifically, we use the implementation of the inverse problem which is provided by OpenAI Gym[2] denoted as *Pendulum-v0*.

The inverted pendulum is a simple 2D-physical simulation. Initially, the pendulum hangs downwards, and has to be swung upwards, by applying force at the joint, either to the left or right. Once the pendulum is oriented upwards, it has to be balanced there as long as possible. This situation is shown in Fig. 11.1.

The state of this problem's environment is composed of three values: the sine and cosine of the pendulum angle, and the angular velocity. The action is the applied torque (with a sign representing a change of direction), and the reward is computed with $-(\text{angle}^2 + 0.1 * \text{velocity}^2 + 0.001 * \text{torque}^2)$. This ensures that the learning process sees the largest rewards if the pendulum is upright (angles are close to zero), moving slowly (small angular velocities), with little effort (small torques).

---

[2] OpenAI Gym is a collection of reinforcement learning problems; see https://github.com/openai/gym for details such as installation instructions.

11.2.3 Learning Algorithm

Largely, we leave the algorithm used in the original Keras tutorial as is (Singh 2020). In fact, this algorithm follows the concept of the Deep Deterministic Policy Gradient (DDPG) algorithm by Lillicrap et al. (2015). We will not go into all details of the algorithm, but will note some important aspects for setting up the tuning procedure.

The learning algorithm essentially uses four different networks: an actor network, a target actor network, a critic network, and a target critic network. The actor network represents the policy of the agent: mapping from states to actions. The critic network tries to guess the value (in terms of future rewards) of the current state/action pair, thus providing a baseline to compare the actor against. That is, the critic network maps from states *and* actions to a kind of estimated reward value.

The respective target networks are copies of these two networks. They use the same architecture and weights, which are not directly trained for the target networks but are instead updated via cloning them from the original networks regularly during the learning process. These concepts (the actor-critic concept and target networks) are intended to stabilize network training.

The learning algorithm also makes use of *experience replay*, which represents a collection (or buffer) of tuples consisting of states, actions, rewards, and new states. This allows learning from a set of previously experienced agent-environment interactions, rather than just updating the model with the most recent ones.

11.3 Setting up the Tuning Experiment

11.3.1 File: run.py

The learning algorithm and task environment are processed with PYTHON code, in the file run.py. This is to a large extent identical to the Keras tutorial (Singh 2020).

Here, we explain the relevant changes, showing some snippets from the code.

- The complete code is wrapped into a function, which will later be called from R via the reticulate interface.

```
def run_ddpg(num_hidden,critic_lr,actor_lr,
    gamma,tau,activation,max_episodes,seed):
```

Importantly, the arguments consist of the parameters that will be tuned, as well as max_episodes (number of learning episodes that will be run) and a seed for the random number generator.

- Respectively, these parameters have all been changed from the original, hard-coded values in the Keras tutorial. The original (default) values in the tutorial are num_hidden=256, critic_lr=0.002, actor_lr=0.001, gamma=0.99, tau=0.005, activation = "relu".

- Note that we vary only the size of the largest layers in the networks (default: 256). Especially, the critic has smaller layers that collect the respective inputs (states, actions). These remain unchanged.
- Via the argument `activation`, we only replace the activation functions of the internal layers, not the activation function of the final output layers.
- To make sure that results are reproducible, we set the random number generator seeds (for the reinforcement learning environment, TensorFlow, and NumPy):

```
env.seed(seed)
tf.random.set_seed(seed)
np.random.seed(seed)
```

- We remove the plots from the tutorial code, as these are not particularly useful during automated tuning.
- Finally, we return the variable avg_reward_list, which is the average reward of the last 40 episodes. This returned value will be the objective function value that our tuner Sequential Parameter Optimization Toolbox (SPOT) observes.
- Note that all reward values we consider will be negated, since most of the procedures we employ assume smaller values to be better.

More details on the tuned parameters are given next.

11.3.2 Tuned Parameters

In the previous Sect. 11.3.1, we already briefly introduced the tuned parameters and their default values: num_hidden=256, critic_lr=0.002, actor_lr=0.001, gamma=0.99, tau=0.005, activation="relu". Some of these we may recognize, matching parameters of neural networks that we considered throughout other parts of this book: num_hidden corresponds to the previously discussed `units`, but is a scalar value (it is reused to define the size of all the larger layers in all networks). Instead of a single `learning_rate`, we have separate learning rates for the actor and critic networks, critic_lr, and actor_lr.

The parameter gamma is new, as it is specific to actor-critic learning algorithms: it represents a discount factor which is applied to estimated rewards as a multiplicator. The parameter tau is also new, representing a multiplicator that is used when updating the weights of the target networks. Finally, `activation` is the activation function (here: shared between all internal layers). The parameters and their bounds are summarized in Table 11.1.

List of configurations

The following code snippet shows the code used to define this parameter search space for SPOT in R.

Table 11.1 The hyperparameters for our reinforcement learning example. Note that the defaults and bounds concern the actual scale of each parameter (not transformed). Defaults denote the values from the original Keras tutorial, not the formal defaults from the Keras function interfaces

Name	Type	Default	Scale transformation	Lower bound	Upper bound
num_hidden	Integer	x	256	8	256
critic_lr	Double	10^x	0.002	$1e-5$	$1e-1$
actor_lr	Double	10^x	0.001	$1e-5$	$1e-1$
gamma	Double	$1-10^x$	0.99	0.5	1
tau	Double	10^x	0.001	$1e-4$	$1e-0$
activation	Factor	relu	relu, swish, sigmoid		

```
## configuration for the tuning problem
cfg <- list(
  ## Names of the parameters
  tunepars = c("num_hidden","actor_lr","critic_lr",
               "gamma","tau","activation"),
  ## their lower bounds
  lower =    c(8,         -5,          -5,          -4,       -4,  1),
  ## their upper bounds
  upper =    c(256,       -1,          -1,          -0.3,      0,  3),
  ## their type
  type =     c("integer","numeric","numeric","numeric","numeric","factor"),
  ## transformations to apply
  transformations =
    c(trans_id,trans_10pow,trans_10pow,
      trans_1minus10pow,trans_10pow,trans_id),
  ## another parameter that will not be tuned, but is fixed
  fixpars = list(max_episodes=50L),
  ## specify levels of categorical parameters
  ## (i.e., to translate from integers to these factor levels):
  factorlevels = list(activation=c("relu","swish","sigmoid")),
  ## not used in this example
  ## (specify parameters that are relative to other parameters)
  relpars = list()
)
```

Note that we set a single fixed parameter, max_episodes, limiting the evaluation of the learning process to 50 episodes.

11.3.3 Further Configuration of SPOT

SPOT is configured to use 300 evaluations, which are spent as follows: Each evaluation is replicated (evaluated repeatedly) five times, to account for noise. Noise is a substantial issue in reinforcement learning cases like this one.

30 different configurations are tested in the initial design, leading to 150 evaluations (including the replications). The remaining 150 evaluations are spent by the iterative search procedure of SPOT. Due to replications, this implies that 30 further configurations are tested. Also, due to the stochastic nature of the problem, we set the parameter `noise=TRUE`.

The employed surrogate model is Kriging (a.k.a. Gaussian process regression), which is configured to use the so-called nugget effect (`useLambda=TRUE`), but no re-interpolation (`reinterpolate=FALSE`).

In each iteration after the initial design, a Differential Evolution algorithm is used to search the surrogate model for a new, promising candidate. The Differential Evolution algorithm is allowed to spend 2400 evaluations of the surrogate model in each iteration of SPOT.

For the sake of reproducibility, random number generator seeds are specified (`seedSPOT, seedFun`). Each replication will work with a different random number generator seed (iterated, starting from `seedFun`).

Arguments for calling SPOT

The respective configuration and function call is

```
result <- spot(fun = objf,
               lower=cfg$lower,
               upper=cfg$upper,
               control = list(types=cfg$type,
                              funEvals=300,
                              plots=TRUE,
                              optimizer=optimDE,
                              noise=TRUE,
                              seedSPOT=1,
                              seedFun=1,
                              designControl=list(size=5*length(cfg$lower),
                                                 replicates=5),
                              replicates=5,
                              model=buildKriging,
                              modelControl=list(target="ei",useLambda=TRUE,
                                                reinterpolate=FALSE),
                              optimizerControl=list(funEvals=
                                                400*length(cfg$lower))
                              )
               )
```

11.3.4 Post-processing and Validating the Results

To determine how well the tuning worked, we perform a short validation experiment at the end. There, we spend 10 replications to evaluate the best found solution. We also spend more episodes for this test (i.e., max_episodes=100). This provides a less noisy and more reliable estimate of our solution's performance, compared to the respective performance of the default settings from the tutorial (see Table 11.1).

Note that this step requires a bit of data processing, where we first aggregate our result data set by computing mean objective values (i.e., over the 5 replications), to determine which configuration was evaluated to work best on average.

11.4 Results

Table 11.2 compares the parameters of the best solution found during tuning with those of the defaults from the tutorial. It also lists the respective performance (average reward) and its standard deviation. We can load the result file created after tuning to create a visual impression of this comparison (Fig. 11.2).

```
load("supplementary/ch11-caseStudyIV/resultFile.RData")
boxplot(best_real_y,default_y,
        names=c("tuned","default"),
        xlab="performance (-reward)",
        horizontal=TRUE)
```

Interestingly, much smaller size of the dense layers (num_hidden=64) seems to suffice for the tuned solution. The larger tutorial network uses 256 units. The tuned algorithm also uses a larger learning rate for the critic network, compared to the actor network. The parameters gamma and tau deviate strongly from the respective defaults.

Table 11.2 The hyperparameter values of the best solution found during tuning, compared against those of the defaults from the Keras tutorial by Singh (2020). It also lists the respective performance (mean neg. reward) and its standard deviation. Mean and standard deviation are computed over 10 replications, evaluated with 100 episodes

Variable name	Default	Tuned
num_hidden	256	64
critic_lr	0.00200	0.00349
actor_lr	0.00100	0.00074
gamma	0.99000	0.93668
tau	0.00100	0.01481
activation	relu	swish
Average negated reward	183.62	169.86
st. dev. of avg. neg. reward	37.49	27.60

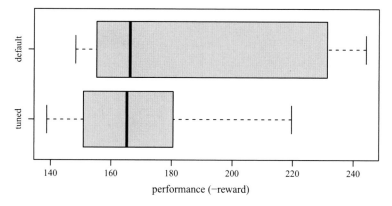

Fig. 11.2 Boxplot comparing the default and tuned configurations of the reinforcement learning process, in terms of their negated reward

11.5 Severity: Validating the Results

Let us proceed to analyze the average negated reward attained between the tuned and default parameters using severity. The pre-experimental runs indicate that the difference is $\bar{x} = 13.76$. Because this value is positive, we can assume that the tuned solution is superior. The standard deviation is $s_d = 32.67$. Based on Eq. 5.14, and with $\alpha = 0.05$, $\beta = 0.2$, and $\Delta = 40$, we can determine the number of runs for the full experiment.

For a relevant difference of 40, approximately 8 completing runs per algorithm are required. Hence, we can proceed directly to evaluate the severity as sufficient runs have already been performed.

The decision based on the p-value of 0.0915 is to not reject H_0. Considering a target relevant difference $\Delta = 40$, the severity of not rejecting H_0 is 0.99, and thus it strongly supports the decision of not rejecting the H_0. The corresponding severity plot is shown in Fig. 11.3. Analyzing the results of hypothesis testing and severity as shown in Table 11.3, the differences in terms of parameter values do not seem to manifest in the performance values. It can be observed in Table 11.2 that a comparatively minor difference in mean performance is observed, while the difference in standard deviation is a bit more pronounced. However, this cannot be deemed as statistically significant relevance.

Overall, this matches well with what we see from a more detailed look at the tuning results (Fig. 11.4):

```
SPOTMisc::plot_parallel(resultpp)
```

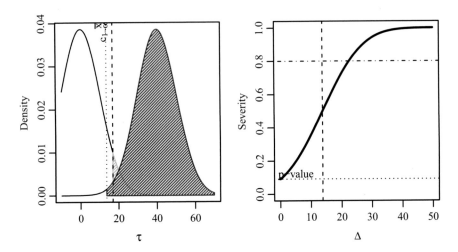

Fig. 11.3 Reinforcement learning. Severity of not rejecting H0 (red), power (blue), and error (gray). Left: the observed mean $\bar{x} = 13.76$ is smaller than the cut-off point $c_{1-\alpha} = 16.99$. Right: Severity of not rejecting H0 as a function of Δ

Table 11.3 Case Study IV: Result Analysis

p-value	Decision	Power	Cohen's d	Hedge's g	Severity
0.09	H0 not rejected	0.987019	0.4178644	0.4002082	$\Delta \geq 25$ are well supported

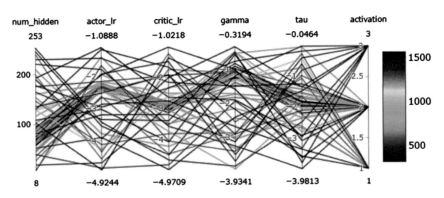

Fig. 11.4 Parallel plot of the achieved performance values during tuning. Red lines denote poor configurations (poor performance), blue lines better configurations

11.6 Summary and Discussion

In summary, the investigation shows that large parts of the search space lead to poorly performing configurations of the algorithm. Still, there seems to be a broad spectrum of potentially well-performing configurations, which interestingly include fairly small networks (i.e., with few units). This observation may be linked to the complexity of the problem, which is a relatively simple reinforcement learning scenario with few states and actions.

The tuned solution seems to work a little better than the default settings from the tutorial (Singh 2020), but those defaults are still competitive. It is reasonable to assume that the tutorial defaults were chosen with care (potentially by some sort of tuning procedure, or relevant experience by the tutorial's author) and are well-suited for this problem. While the smaller network implies faster computation, the larger network has the advantage of being more easily transferred to more complex reinforcement learning cases.

Chapter 12
Global Study: Influence of Tuning

Martin Zaefferer, Olaf Mersmann, and Thomas Bartz-Beielstein

Abstract Expanding the more focused analyses from previous chapters, this chapter takes a broader view at the tuning process. That means, rather than tuning an individual model, this investigation considers the tuning of multiple models, with different tuners, and varying data sets. The core aim is to see how characteristics of the data and the model choice may impact the tuning procedure. We investigate five hypotheses, concerning the necessity of tuning, the impact of data characteristics, the impact of the target variable type, the impact of model choice, and benchmarking. Not only does this entail an in-depth tuning study, but we also tie our results to a measure of problem difficulty and use consensus ranking to aggregate the diverse experimental results.

12.1 Introduction

As described in Sect. 4.4, the Sequential Parameter Optimization Toolbox (SPOT) offers a robust approach for the tuning of Machine Learning (ML) algorithms, especially if the training and/or evaluation run time become large.

In practice, the learning process of models, $\mathcal{A}$, and hence the choice of their hyperparameters, λ, is influenced by a plethora of other factors. On the one hand, this complex situation further motivates the use of tuning procedures, since the ML algorithms have to be adapted to new data or situations. On the other hand, this

M. Zaefferer (✉)
Bartz & Bartz GmbH and with Institute for Data Science, Engineering, and Analytics, TH Köln, Gummersbach, Germany

Duale Hochschule Baden-Württemberg Ravensburg, Ravensburg, Germany
e-mail: zaefferer@dhbw-ravensburg.de

O. Mersmann · T. Bartz-Beielstein
Institute for Data Science, Engineering, and Analytics, TH Köln, Steinmüllerallee 1, 51643 Gummersbach, Germany
e-mail: olaf.mersmann@th-koeln.de

T. Bartz-Beielstein
e-mail: thomas.bartz-beielstein@th-koeln.de

© The Author(s) 2023
E. Bartz et al. (eds.), *Hyperparameter Tuning for Machine and Deep Learning with R*,
https://doi.org/10.1007/978-981-19-5170-1_12

raises the question of how such factors influence the tuning process itself. We want to investigate this in a structured manner.

In detail, the following factors, which are introduced in Sect. 8.2.1, are objects of our investigation.[1] We test their influence on the tuning procedures.

- Number of numerical features in the data (nnumericals).
- Number of categorical features in the data (nfactors).
- Cardinality of the categorical features, i.e., the maximal number of levels (cardinality).
- Number of observations in the data (n).
- Task type, classification or regression.
- Choice of model.

12.2 Research Questions

We want to investigate the following hypotheses.

(H-1)   Tuning is necessary to find good parameter values (compared to defaults).
(H-2)   Data: Properties of the data influence (the difficulty of) the tuning process.

- Information content: If the data has little information content, models are easier to tune, since more parameter configurations achieve near-optimal quality. In general, changing parameters has less impact on model quality in this case.
- Number of features: A larger number of features leads to longer run times, which affects how many evaluations can be made during tuning.
- Type of features: The number of numerical and/or categorical features and their cardinality influences how much information is available to the model, hence may affect the difficulty of tuning.
- Number of observations n: With increasing n, the average run time of model evaluations will increase.

(H-3)   Target variable: There is no fundamental difference between tuning regression or classification models.
(H-4)   Model: The choice of model (e.g., Elastic Net (EN) or Support Vector Machine (SVM)) affects the difficulty of the tuning task, but not necessarily the choice of tuning procedure.
(H-5)   Benchmark: The performance of the employed tuners can be measured in a statistically sound manner.

---

[1] The data from Census-Income (KDD) Data Set (CID) is historical. It includes wording or categories regarding people which do not represent or reflect any views of the authors and editors.

12.3 Setup

To investigate our research questions, the data set needs to be pre-processed accordingly. This pre-processing is documented in the included source code and is explained in Sect. 8.2.1.

For classification, the experiments summarized in Table 12.1 are performed with each tuner and each model. A reduced number of experiments are performed for regression, see Table 12.2.

To judge the impact of creating random subsets of the data (i.e., to reduce n), and to consider the test/train split, three data sets are generated for each configuration. All experiments are repeated for each of those three data sets.

Table 12.1 Experiments for classification: Investigated combinations of the number of categorical features (nfactors), numerical features(nnumericals), cardinality, and n. An empty field for cardinality occurs for low nfactors. In that case, no categorical features are present, so the number of categories becomes irrelevant. The number of observations, n, is varied on a logarithmic scale with five levels in the range from 10^4 to 10^5, i.e., 10 000, 17 783, 31 623, 56 234, and 100 000

nfactors	nnumericals	Cardinality	Number of observations (n)
High	Low	Low	$10^4, 10^{4.25}, \ldots, 10^5$
Medium	Medium	Low	$10^4, 10^{4.25}, \ldots, 10^5$
Low	High		$10^4, 10^{4.25}, \ldots, 10^5$
High	High	Low	$10^4, 10^{4.25}, \ldots, 10^5$
High	Low	Medium	$10^4, 10^{4.25}, \ldots, 10^5$
Medium	Medium	Medium	$10^4, 10^{4.25}, \ldots, 10^5$
High	High	Medium	$10^4, 10^{4.25}, \ldots, 10^5$
High	Low	High	$10^4, 10^{4.25}, \ldots, 10^5$
Medium	Medium	High	$10^4, 10^{4.25}, \ldots, 10^5$
High	High	High	$10^4, 10^{4.25}, \ldots, 10^5$
Complete data set			299285

Table 12.2 Experiments for regression: Investigated combinations of the number of categorical features (nfactors), numerical features (nnumericals), cardinality, and n. An empty field for cardinality occurs for low nfactors. In that case, no categorical features are present, so the number of categories becomes irrelevant

nfactors	nnumericals	Cardinality	Number of observations (n)
low	High		$10^4, 10^{4.25}, \ldots, 10^5$
High	High	Low	$10^4, 10^{4.25}, \ldots, 10^5$
High	High	High	$10^4, 10^{4.25}, \ldots, 10^5$

12.3.1 Model Configuration

The experiments include the models k-Nearest-Neighbor (KNN), Decision Tree (DT), EN, Random Forest (RF), Gradient Boosting (GB), and SVM. Their respective parameters are listed in Table 4.1.

For EN, `lambda` is not optimized by the tuner. Rather, the `glmnet` implementation itself tunes that parameter. Here, a sequence of different `lambda` values is tested (Hastie and Qian 2016; Friedman et al. 2020).

For SVM, the choice of the kernel (`kernel`) is limited to `radial` and `sigmoid`, since we experienced surprisingly large runtimes for `linear` and `polynomial` in a set of preliminary experiments. Hence, `degree` is also excluded, as it is only relevant for the `polynomial` kernel. Due to experiment runtime, we also did not perform experiments with SVM and KNN on data sets with $m \geq 10^5$ observations. These would require using model variants that are able to deal with huge data sets (e.g., some sparse SVM type).

Table 3.8 lists hyperparameters that are actually tuned in the experiments, including data type, bounds, and employed transformations. Here, we mostly follow the bounds and transformations as used by Probst et al. (2019a). Fundamentally, these are not general suggestions. Rather, reasonable bounds will usually require some considerations with respect to data understanding, modeling/analysis, and computational resources. Bounds on values which affect run time should be chosen so that experiments are still possible within a reasonable time frame. Similar consideration can apply to memory requirements. Where increasing/decreasing parameters may lead to increasing/decreasing sensitivity of the model, a suitable transformation (e.g., log-transformation) should be applied.

Most other configurations of the investigated models remain at default values. The only exceptions are:

- `ranger`: For the sake of comparability with other models, model training and evaluation are performed in a single thread, without parallelization (`num.threads = 1`).
- `xgboost`: Similarly to `ranger`, we set `nthread=1`. For regression, the evaluation metric is set to the root-mean-square error (`eval_metric="rmse"`). For classification, log-loss is chosen (`eval_metric="logloss"`).

12.3.2 Runtime For the Global Study

Similar to the local studies (Chaps. 8–10), run times are recorded in addition to Mean Mis-Classification Error (MMCE) (MMCE is defined in Eq. (2.2)). The recorded run times are overall run time of a model evaluation, run time for prediction, and run time for training.

Runtime budget: To mirror a realistic use case, we specify a fixed run time budget for the tuner. This limits how long the tuner may take to find potentially optimal hyperparameter values. We set a budget of 5 h for SVM, KNN, RF, and GB. Since EN and DT are much faster to evaluate and less complex, they receive a considerably lower budget (EN: 1 h, DT: 5 min).

Timeout: For a majority of the models, the run time of a single evaluation (training + prediction) is hard to predict and may easily become excessive if parameters are chosen poorly. In extreme cases, the run time of a single evaluation may become so large that it consumes the bulk of the tuner's allotted run time, or more. In such a case, there would be insufficient time to test different hyperparameter values. To prevent this, we specify a limit for the run time of a single evaluation, which we call timeout. If the timeout is exceeded by the model, the evaluation will be aborted. During the experiments, we set the timeout to a twentieth of the tuner's overall run time budget. Exceptions are the experiments with DT (`rpart`): Since `rpart` evaluates extremely quickly, (in our experiments: usually much less than a second) and has a correspondingly reduced run time budget (5 min), the timeout is not required. In fact, tracking the timeout would add considerable overhead to the evaluation time in this case.

Additional performance measure from the `mlr` package could be easily integrated when necessary.[2]

12.3.2.1 Surrogate Model

In one important case, we deviate from more common configurations of surrogate model based optimization algorithms: For the determination of the next candidate solution to be evaluated, we directly use the predicted value of the Gaussian process model, instead of the so-called expected improvement. Our reason is, that the expected improvement may yield worse results if the number of evaluations is low, or the dimensionality rather high (Rehbach et al. 2020; Wessing and Preuss 2017). With the strictly limited run time budget, our experience is that the predicted value is preferable. A similar observation is made by De Ath et al. (2019).

12.3.2.2 Configuration of Random Search

With Random Search (RS), hyperparameter values will be sampled uniformly from the search space. All related configurations (timeout, run time budget, etc.) correspond to those of SPOT.

---

[2] A complete list of measures in `mlr` can be found at https://mlr.mlr-org.com/articles/tutorial/measures.html.

12.3.3 Benchmark of Runtime

The experiments are not run on entirely identical hardware, but on somewhat diverse nodes of a cluster. Hence, we have to consider that single nodes are faster or slower than others (i.e., a tuning run on one node may perform more model evaluations than on another node, simply because of having more computational power). To preserve some sort of comparability, we compute a corrective factor that will be multiplied with the run time budget. Before each tuning run, we compute a short performance benchmark. The time measured for that benchmark will be divided by the respective value measured at a reference node, to determine the run time multiplicator. We use the `benchmark_std` function from the `benchmarkme` R package (Gillespie 2021).

12.3.4 Defaults and Replications

As a comparison basis, we perform an additional experiment for each model, where all hyperparameter values remain at the models default settings. However, in those cases we do *not* set a timeout for evaluation. Since no search takes place, the overall run time for default values is anyways considerably lower than the run time of SPOT or RS. All other settings correspond to those of SPOT and RS. To roughly estimate the variance of results, we repeat all experiments (each tuner for each model on each data set) three times.

12.4 Results

In this section, we provide an exploratory overview of the results of the experiments. A detailed discussion of the results in terms of the research questions defined in Sect. 12.2 follows in Sect. 12.5.

To visualize the overall results, we show exemplary boxplots.[3] Since different results are achieved depending on the data set and optimized model, a preprocessing step is performed first: For each sampled data set and each model, the mean value of all observed results (model quality of the best solutions found) is determined. This mean is then subtracted from each individual observed value of the corresponding group. Subsequently, these subtracted individual values are examined. This allows a better visualization of the difference between the tuners without compromising interpretability. The resulting values are no longer on the original scale of the model quality, but the units remain unchanged. Thus, differences on this scale can still be interpreted well.

---

[3] Corresponding boxplots for all experiments can be found in the appendix of this document.

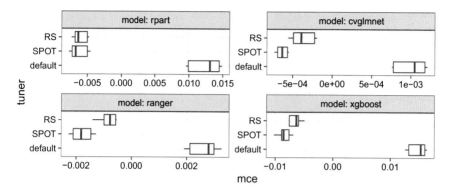

Fig. 12.1 Boxplot comparing tuners for different classification models, with $m = 10^5$, nfactors = high, nnumericals = high, cardinality = high. For this value of n, no experiments with KNN and SVM are performed. The presented quality measure (MMCE) shows values after preprocessing, where the mean value for each problem instance is subtracted

For the classification experiments, Fig. 12.1 shows the results for a case with many features and observations (nfactors, nnumericals, and cardinality are all set to high and $m = 10^5$). The figure first shows that both tuners (RS, SPOT) achieve a significant improvement over default values. The value of this improvement is in the range of about 1% MMCE. In almost all cases, the tuners show a smaller dispersion of the quality values than a model with default values. Except for the tuning of rpart, the results obtained by SPOT are better than those of RS.

Likewise for classification, Fig. 12.2 shows the case with $n = 10^4$, without categorical features (nfactors=low) and with maximum number of numerical features (nnumericals=high). Here the data set contains much less information, since a large part of the features is missing and only few observations are available. Still, both tuners are significantly better than using default values. In this case, it is mostly not clear which of the two tuners provides better results. For rpart (DT) and SVM, RS seems to work better.

In the same form as for classification, Figs. 12.3, and 12.4 show results for regression models. Unlike for classification, the results here are somewhat more diverse. In particular, glmnet shows a small difference between the tuners and default values. There are also differences of several orders of magnitude between the individual models (e.g. RF and SVM). For example, for RF the differences between tuners and default values are about 0.02 years (the target variable scale is age in years). As shown in Figs. 12.3 and 12.4, the interquartile range is about 0.01 years.

For GB, on the other hand, there is a difference of about 20 years between tuners and default values. Here, the default values seem to be particularly poorly suited.

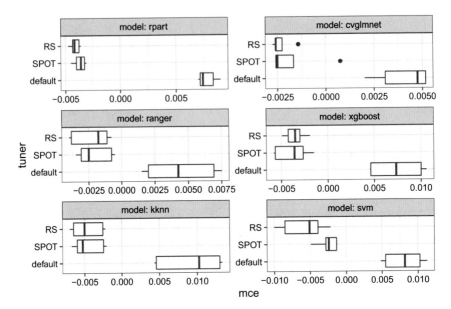

Fig. 12.2 Boxplot comparing tuners for different classification models, with $m = 10^4$, nfactors = low, nnumericals = high. The presented quality measure (MMCE) shows values after preprocessing, where the mean value for each problem instance is subtracted

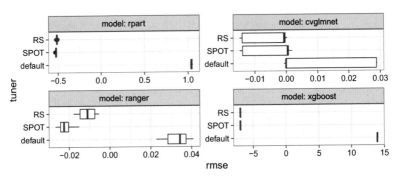

Fig. 12.3 Boxplot comparing tuners for different regression models, with $m = 10^5$, nfactors = high, nnumericals = high, cardinality = high. For this value of n no experiments with KNN and SVM are performed. The presented quality measure (MMCE) shows values after preprocessing, where the mean value for each problem instance is subtracted

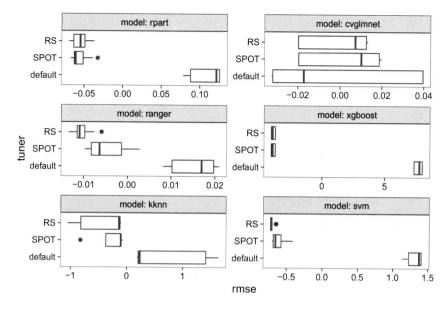

Fig. 12.4 Boxplot comparing tuners for different regression models, with $m = 10^4$, nfactors = low, nnumericals = high. The presented quality measure (MMCE) shows values after preprocessing, where the mean value for each problem instance is subtracted

12.5   Discussion

12.5.1   Rank-Analysis

We analyze the results of the experiments using rankings of the result values instead of the raw results. Rankings are scale invariant, so aggregating the results from different problem instances (resulting from data set and model choice) with different scales for the result values is possible. This is important because we have to aggregate over very diverse problem instances.

To aggregate rankings (also called consensus rankings), we follow the so-called "optimal ranking" approach of Kemeny (1959). Here, the consensus ranking is determined such that the mean distance between the consensus ranking and observed ranking is minimal. The distance measure used for this purpose is Kendall's tau (Kendall 1938), which counts the number of pairwise comparisons in which two rankings contradict each other. The ranking determined using this approach can be interpreted as the "median" of the individual rankings.

This procedure has the following advantages (Hornik and Meyer 2007; Mersmann et al. 2010a):

- Scale invariant.
- Invariant to irrelevant alternatives.

- Aggregation of large sets of comparisons, over arbitrary factors.
- Easy/intuitive to interpret results and visualizable.
- Generates relevant information: selection of the best one.
- Additional weights for preferences can be inserted.
- Fast evaluation.
- Non-parametric method, no distribution assumptions.
- Visualization of clusters over distance is possible, identification of problem classes with similar algorithm behavior.

However, in addition to these advantages, there are also disadvantages:

- Estimating uncertainty in the ranking is difficult.
- The ranking does not have to induce a strict ordering, ties are possible.

We generate the consensus ranking by combining rankings of tuners (SPOT, RS, default) of individual experiments. We always summarize the rankings of 9 experiments (3 repetitions of the tuner runs on each of 3 randomly drawn data sets). Then, to aggregate across different variables related to the study subjects (e.g. n, nfactors), we count how often the tuners reach a certain consensus rank.

This count is divided by the total number of experiments (or number of rankings) on the corresponding problem instances. Thus, we record for each tuner how often a particular consensus rank is achieved.

Simplified example:

- For case 2 (i.e., for a fixed choice of nnumericals, nfactors, cardinality, model, n, target variable), the comparison of SPOT, RS, and default methods resulted in the ranks

$$\{1, 3, 2\} \quad \{1, 2, 3\} \quad \{2, 1, 3\}.$$

The consensus ranking for this case is $\{1, 2, 3\}$.
- For case 2 (i.e., for *another* fixed choice of nnumericals, nfactors, cardinality, model, n, target variable), the comparison of SPOT, RS, and default methods resulted in the ranks

$$\{3, 2, 1\} \quad \{1, 2, 3\} \quad \{2, 1, 3\}.$$

The consensus ranking for this case is $\{2, 1, 3\}$.
- When both experiments are combined for an analysis, the frequencies for the obtained rankings are as follows:

  - SPOT: rank 1 with 50%, rank 2 with 50%, rank 3 with 0%.
  - RS: rank 1 with 50%, rank 2 with 50%, rank 3 with 0%.
  - Default: rank 1 with 0%, rank 2 with 0%, rank 3 with 100%.

12.5.2   Rank-Analysis: Classification

Based on the analysis method described in Sect. 12.5.1, Fig. 12.5 shows the relationship between the tuners, the number of observations n, and the optimized models. It shows that SPOT and RS mostly beat the default setting and SPOT also usually performs better than RS. However, some cases deviate from this. Especially, `glmnet` (EN) and `rpart` (DT) seem to profit less from tuning: here the distinction between the ranks of tuners is more difficult. In addition, when the number of observations is small, there tends to be a greater uncertainty in the results. These results can be partly explained by the required runtime of the models. With a smaller number of observations, the runtime of the individual models decreases, and `glmnet` and `rpart` are the models with the lowest runtime (in the range of a few seconds, or below one second in the case of `rpart`). If the evaluation of the models itself takes hardly any time, it is advantageous for RS that its runtime overhead is low. SPOT, on the other hand, requires a larger overhead (for the surrogate model and the corresponding search for new parameter configurations).

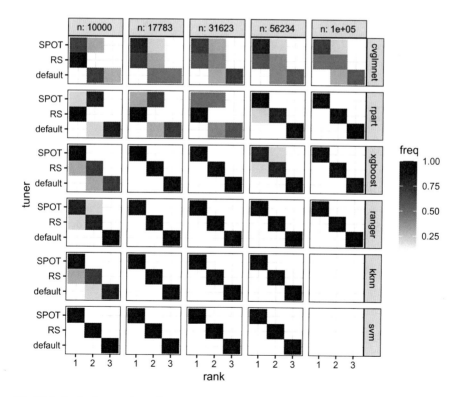

Fig. 12.5 Rank of tuners depending on number of observations (n) and model, for classification

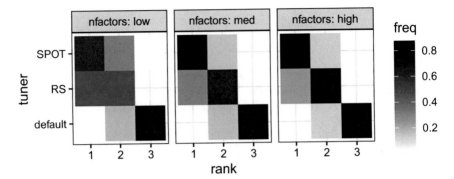

Fig. 12.6 Rank of tuners as a function of the number of categorical features (nfactors) and the model, for classification

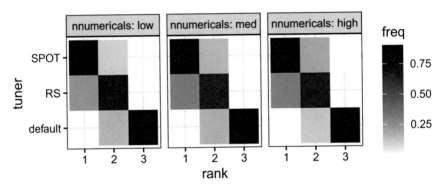

Fig. 12.7 Rank of tuners depending on number of numerical features (nnumericals) and model, for classification

Figure 12.6 shows the corresponding summary of the results depending on the number of categorical features (nfactors). Again, SPOT usually performs best, followed by RS. There is a tendency for the greatest uncertainty to be found in cases where number of categorical features is low. Two explanations are possible: On the one hand, the reduction of features also means a reduction of the required runtime. On the other hand, the difficulty of the modeling increases, since with fewer features less information is available to separate the classes.

The corresponding results for the number of numerical features can be found in Fig. 12.7. There are hardly any differences, the number of numerical features seems to have little influence. It should be noted that the data set contains fewer numerical than categorical features anyway.

The cardinality of the categorical features also has little influence, see Fig. 12.8. However, there is a slight tendency: at higher cardinality, the distinction between the first rank (SPOT) and second rank (RS) is clearer. This can be explained (similarly to nfactors) by the higher information content of the data set.

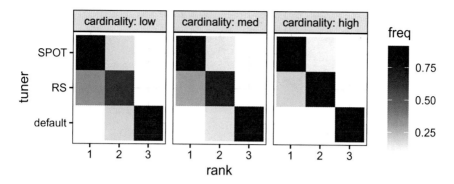

Fig. 12.8 Rank of tuners as a function of the cardinality of categorical features and the model, for classification. *Note* This figure does not include the cases where the data set no longer contains categorical features (cardinality cannot be determined)

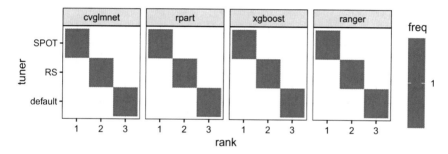

Fig. 12.9 Rank of tuners depending on the model, for classification on the complete data set. Due to the increased runtime, `kknn` (KNN) and `e1071` (SVM) are not included

Finally, Fig. 12.9 shows the result of the tuners on the unmodified, complete data set. For each case, SPOT gets rank 1 and RS rank 2. This result is in line with the trends described above, since the complete data set contains the most information and also leads to the largest runtimes (for model evaluations).

12.5.3 Rank-Analysis: Regression

In addition to the classification experiments, a smaller set of experiments with regression as an objective are also conducted. Figure 12.10 shows the results for this case separately for each optimized model. Here, too, SPOT is usually ranked first. Unlike in case of classification, however, there is more uncertainty.

For `glmnet` (EN), default values occasionally even achieve rank 1. It seems that linear modeling with `glmnet` is unsuitable for regression with the present data set,

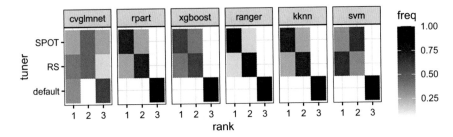

Fig. 12.10 Rank of tuners depending on model for regression

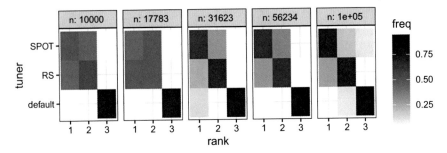

Fig. 12.11 Rank of tuners, depending on number of observations (n), for regression

so that the tuners can hardly achieve differences to the default values. This behavior was already indicated in Fig. 12.4.

In the case of SVM, RS is more often ranked first than SPOT. The reason for this is not clear. A possible cause is a lower influence of the hyperparameters on the model quality of SVM for regression (compared to classification). However, it should also be considered that less experiments were conducted for regression than for classification.

The dependence on the number of observations n is shown in Fig. 12.11. The results for regression with respect to n are largely consistent with those for classification. With an increasing number of observations, SPOT is more clearly ahead of RS.

The correlation with categorical features (number, cardinality) is also consistent with the classification results, see Fig. 12.12. With larger number of features and larger cardinality, a clearer separation between the tuners is observed.

12.5.4 Problem Analysis: Difficulty

In the context of this study, an interesting question arose as to how the difficulty of the modeling problem is related to the results of the tuning procedures. We investigate this on the basis of the data obtained from the experiments.

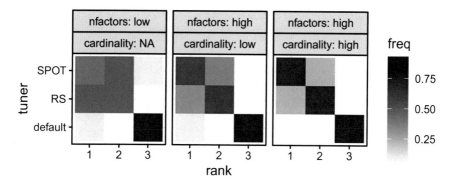

Fig. 12.12 Rank of tuners, depending on cardinality and number of categorical features (nfactors), for regression

In general, there are many measures of the difficulty of modeling problems in the literature. An overview is given by Lorena et al. (2019). Of these measures, the overlap volume (F2, see Lorena et al. 2019) is of interest, since it is easily interpretable and not specific to a particular model.

First, the values of each feature are separated into two groups, one for each class. Then, the range of the intersection of the two groups is computed for each individual feature. This is called the *overlap* of a feature. The *overlap volume* of the data set is then the product of the individual overlap values.

However, this measure is unsuitable for categorical features (even after dummy coding). Furthermore, outliers are very problematic for this measure.

We therefore use a slight modification: We calculate the proportion of sample values for each feature, which could occur in both classes (i.e. for which a swap of the classes based on the feature value is possible). As an example, this is illustrated for a numeric and a categorical feature in Fig. 12.13. For the overall data set, the individual overlap values of each feature are multiplied. Subsequently, we refer to this measure as *sample overlap*.

Figure 12.14 shows the dependence of the sample overlap on our data properties (n, nfactors, nnumericals, cardinality).

Our data sets can be grouped into 4 difficulty levels based on these values of sample overlap:

1. Sample overlap ≈ 0.39: nfactors = high and cardinality = high (green).
2. Sample overlap ≈ 0.54: nfactors = high and cardinality = med (blue).
3. Sample overlap ≈ 0.76: all others (orange).
4. Sample overlap ≈ 1.00: nfactors = low (red).

Here, 4 corresponds to the highest level of difficulty. For the range relevant in the experiments, there is almost no change depending on n or nnumericals. For nfactors and cardinality a strong correlation can be seen.

Based on the 4 difficulty levels, the ranks already determined in previous sections can be re-ranked. The result is summarized in Fig. 12.15. It turns out that as the

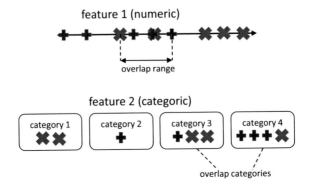

Fig. 12.13 Example of the sample overlap for two features of a classification problem with two classes. The crosses show samples from the data set. All samples with class 1 are red (+). All samples with class 2 are blue (x). For feature 1, the overlap is 50% (number of samples in the overlap area divided by the total number of samples). For feature 2, the overlap is 70% (number of samples in the overlap categories divided by total number of samples). For both characteristics together, the sample overlap is $0.5 \times 0.7 = 0.35$

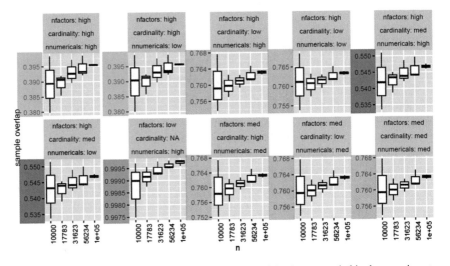

Fig. 12.14 Sample overlap depending on the properties of the data set varied in the experiments. The sample overlap is used as a measure of problem difficulty and leads to the definition of four difficulty levels (marked in color)

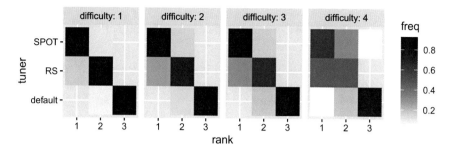

Fig. 12.15 Frequency of the achieved ranks of the tuners depending on the difficulty level

difficulty of the problem increases, the rank differences are less robust. That is, with larger sample overlap it is harder to estimate which tuner works best. This is plausible with respect to the theoretical extreme case: A maximally difficult data set cannot be learned by any model, since the features no longer contain any information about the classes. In this case, the hyperparameters no longer have any influence and the tuners cannot improve.

12.6 Summary and Discussion

To summarize our findings, we discuss the hypotheses from Sect. 12.2.

(H-1) *Tuning is necessary:* In almost all cases, significantly worse results are obtained with default settings. In addition, the variance of the results decreases when tuners are used (see Figs. 12.1, 12.2, 12.3 and 12.4).

(H-2) *Data:* Differences between tuners become apparent for data sets with a high information content and for larger data volumes. If the number of levels or features (categorical and numerical) decreases, it can be observed that the differences between the tuners become smaller.

It has been confirmed that as the number of features and observations increases, the runtime increases. As the mean runtime of the models increases (i.e., for more complex models or larger data sets), SPOT performs increasingly better than RS, as the ratio of overhead to evaluation time decreases more for SPOT.

(H-3) *Target variable:* The execution of the tuning is not affected by the change of the target variable. A peculiarity that requires further investigation occurred when tuning SVM for regression: RS seems to perform better than SPOT. We recommend using a larger data base to investigate this case.

(H-4) *Model:* The choice of tuning method is not fundamentally influenced by the models. Models that can be evaluated very quickly (e.g. `rpart`) benefit from the larger number of evaluations. This is due to the fact that time was chosen as a termination criterion.

(H-5)   *Benchmark:* As described in Sect. 12.5.1, analysis methods based on consensus ranking can be used to evaluate the suitability of tuners in a simple and statistically valid way.

Overall, the result of this experimental investigation underlines that tuning of machine learning models is essential to produce good configurations of the hyperparameters and thus models of high quality.

It is especially important to use tuners at all, since the biggest differences are usually observed in comparison to default settings. However, there are also differences between tuners, which in our investigation are mostly favoring the use of SPOT, i.e. a model-based optimizers. For the prospective focus on tuning with relatively large data sets (large n and k), this is all the more evident, since the resulting high model runtime favors the use of model-based methods.

Since the number of hyperparameters is manageable ($<< 100$), the addition of a few more parameters does not significantly increase the complexity of the tuning. For both Random Search (Bergstra and Bengio 2012) and SPOT, the addition of a few parameters is harmless, even if they have only a small effect on the model quality.

Moreover, the analysis results in additional benefits:

- Possible software bugs can be detected,
- seemingly unimportant parameters are detected as important,
- unknown interactions can be uncovered.

The surrogate model allows to learn the influence of the hyperparameters on the model values. Unimportant parameters are consequently weighted less in the search.

Finally, we recommend for the selection of the tuner, in the context of this study:

- Random Search can be used when either very little time (order of magnitude: time is sufficient for a single-digit number of sequential evaluations) but a lot of parallel computing capacity is available or when models can be evaluated extremely fast (in the range of seconds).
- If time-consuming computations with complex data and models need to be performed in more time (with relatively limited parallel computing capacity), we recommend model-based tuning with SPO.
- In exceptional cases, which deviate from our considered objective of tuning with complex data and models (extremely large amount of computing time available, average evaluation times for the models), the use of surrogate model-free tuning methods may also be considered.

Appendix
Software Installations

A.1 Installing SPOT and SPOTMisc

`SPOT` version 2.11.14 and `SPOTMisc` version 1.19.40, which are available on Comprehensive R Archive Network (CRAN), were used to execute the code in this book. Most recent versions of these packages can be downloaded from https://www.spotseven.de/spot/.

A.2 Installing Python

A.2.1 Create and Activate a PYTHON *Environment in the Project*

It is recommended that one PYTHON virtual environment is used per experiment. Navigate into the project directory by using the following command:

```
cd <project-dir>
```

© The Editor(s) (if applicable) and The Author(s) 2023
E. Bartz et al. (eds.), *Hyperparameter Tuning for Machine and Deep Learning with R*,
https://doi.org/10.1007/978-981-19-5170-1

Create a new virtual environment in a folder called `python` within the project directory using the following command:

```
python3 -m venv python
```

The virtualenv can be activated using the following command in a terminal:

```
source python/bin/activate
```

To verify that the correct version of PYTHON was activated the following command can be executed in a terminal:

```
which python
```

A.2.2 Install PYTHON Packages in the Environment

Python packages such as tensorflow-datasets, numpy, pandas, matplotlib, and other packages can be installed in the PYTHON virtualenv by using `pip install`:

```
python -m pip install -U pip
python -m pip install tensorflow tensorflow-datasets numpy pandas matplotlib
```

A.2.3 Install and Configure Reticulate to Use the Correct PYTHON Version

Install the `reticulate` package using the following command in the R console (e.g., from within RStudio):

```
install.packages("reticulate")
```

To configure `reticulate` to point to the PYTHON executable in the virtualenv `python` from above, create a file in the project directory called `.Rprofile` with the following contents:

```
Sys.setenv(RETICULATE_PYTHON = "../python/bin/python")
```

R software environment for statistical computing and graphics (R) (or RStudio) must be restarted for the setting to take effect. To check that `reticulate` is configured for the correct version of PYTHON the following command can be used in the R (or RStudio) console:

```
reticulate::py_config()
```

A.3   Installing Keras

To get started with Keras, the Keras R package, the core Keras library, and a backend tensor engine (such as TensorFlow (TF)) must be installed. This can be done as follows from within R (or RStudio):

```
install.packages("tensorflow")
install.packages("keras")
library("keras")
install_keras()
```

A.4   Installing SPOT

The following commands can be used to install the most recent version of Sequential Parameter Optimization Toolbox (SPOT) and the additional package Sequential Parameter Optimization Toolbox–Miscelleanous Functions (SPOTMisc) from CRAN:

```
install.packages("SPOT")
install.packages("SPOTMisc")
```

Further information about the most recent SPOT versions will be published on https://www.spotseven.de/spot/.

A.5   Installation on ARM Macs

The following information is based on https://gist.github.com/juliasilge/035d54c594
36604d6142ebebf29aa224.

```
brew install --cask R curl -OL
https://github.com/conda-forge/miniforge/releases/latest/download/Miniforge3-MacOSX-arm64.sh
chmod +x Miniforge3-MacOSX-arm64.sh sh Miniforge3-MacOSX-arm64.sh -b
source ~/miniforge3/bin/activate conda update -y -n base conda conda
create -y --name tf-2.6 python=3.9 conda activate tf-2.6 conda
install -y -c apple tensorflow-deps python -m pip install
tensorflow-macos python -m pip install tensorflow-metal echo
"export
RETICULATE_PYTHON=~/miniforge3/envs/tf-2.6/bin/python" >>
~/.Renviron
```

References

M. Abadi, A. Agarwal, P. Barham, E. Brevdo, Z. Chen, C. Citro, G.S. Corrado, A. Davis, J. Dean, M. Devin, S. Ghemawat, I. Goodfellow, A. Harp, G. Irving, M. Isard, Y. Jia, R. Jozefowicz, L. Kaiser, M. Kudlur, J. Levenberg, D. Mané, R. Monga, S. Moore, D. Murray, C. Olah, M. Schuster, J. Shlens, B. Steiner, I. Sutskever, K. Talwar, P. Tucker, V. Vanhoucke, V. Vasudevan, F. Viégas, O. Vinyals, P. Warden, M. Wattenberg, M. Wicke, Y. Yu, X. Zheng, *TensorFlow: Large-Scale Machine Learning on Heterogeneous Systems* (2015). https://www.tensorflow.org/. Software available from tensorflow.org

M. Abadi, A. Agarwal, P. Barham, E. Brevdo, Z. Chen, C. Citro, G.S. Corrado, A. Davis, J. Dean, M. Devin, S. Ghemawat, I. Goodfellow, A. Harp, G. Irving, M. Isard, Y. Jia, R. Jozefowicz, L. Kaiser, M. Kudlur, J. Levenberg, D. Mane, R. Monga, S. Moore, D. Murray, C. Olah, M. Schuster, J. Shlens, B. Steiner, I. Sutskever, K. Talwar, P. Tucker, V. Vanhoucke, V. Vasudevan, F. Viegas, O. Vinyals, P. Warden, M. Wattenberg, M. Wicke, Y. Yu, X. Zheng, TensorFlow: Large-Scale Machine Learning on Heterogeneous Distributed Systems, March 2016. arXiv:1603.04467

T. Akiba, S. Sano, T. Yanase, T. Ohta, M. Koyama, Optuna: a next-generation hyperparameter optimization framework, in *Proceedings of the 25rd ACM SIGKDD International Conference on Knowledge Discovery and Data Mining* (2019)

H. Arafat Abu Alfeilat, A.B.A. Hassanat, O. Lasassmeh, A.S. Tarawneh, M.B. Alhasanat, H.S. Eyal Salman, V.B. Surya Prasath (2019) Effects of distance measure choice on k-nearest neighbor classifier performance: a review. Big Data **7**(4), 221–248 (2019). https://doi.org/10.1089/big.2018.0175

K.J. Arrow, A difficulty in the concept of social welfare. J. Polit. Econ. **58**(4), 328–346 (1950). http://www.jstor.org/stable/1828886

S. Athey, J. Tibshirani, S. Wager, Generalized random forests. Ann. Stat. **47**, 1148–1178 (2019). ISSN 0090-5364. https://doi.org/10.1214/18-AOS1709

N. Awad, N. Mallik, F. Hutter, *DEHB: Evolutionary Hyberband for Scalable, Robust and Efficient Hyperparameter Optimization* (2021). http://arxiv.org/abs/2105.09821

A. Balaji, A. Allen, *Benchmarking Automatic Machine Learning Frameworks* (2018). http://arxiv.org/abs/1808.06492

P. Balaprakash, M. Salim, T.D. Uram, V. Vishwanath, S.M. Wild, DeepHyper: asynchronous hyperparameter search for deep neural networks, in *25th IEEE International Conference on High Performance Computing (HiPC18)* (IEEE, 2018). https://doi.org/10.1109/hipc.2018.00014

T. Bartz-Beielstein, M. Dröscher, A. Gür, A. Hinterleitner, O. Mersmann, D. Peeva, L. Reese, N. Rehbach, F. Rehbach, A. Sen, A. Subbotin, M. Zaefferer, Resource planning for hospitals under special consideration of the covid-19 pandemic: Optimization and sensitivity analysis, in

Proceedings of the Genetic and Evolutionary Computation Conference Companion, GECCO '21, New York, NY, USA (Association for Computing Machinery, 2021), pp. 293–294. ISBN 9781450383516. https://doi.org/10.1145/3449726.3459473

T. Bartz-Beielstein, *New Experimentalism Applied to Evolutionary Computation*. Ph.D. thesis, Universität Dortmund, Germany, Apr. 2005. http://hdl.handle.net/2003/21461

T. Bartz-Beielstein, *Experimental Research in Evolutionary Computation-The New Experimentalism*. Natural Computing Series. (Springer, Berlin, Heidelberg, New York, 2006), 3-540-32026-1. https://doi.org/10.1007/3-540-32027-X

T. Bartz-Beielstein, How to create generalizable results, in *Springer Handbook of Computational Intelligence*, ed. by J. Kacprzyk, W. Pedrycz (Springer, Berlin, Heidelberg, 2015), pp. 1127–1142. ISBN 978-3-662-43504-5. https://doi.org/10.1007/978-3-662-43505-2_56

T. Bartz-Beielstein, M. Preuss, Automatic and interactive tuning of algorithms, in *Proceedings of the 13th Annual Conference Companion on Genetic and Evolutionary Computation*, GECCO '11, New York, NY, USA (Association for Computing Machinery, 2011) pp. 1361–1380. ISBN 9781450306904. https://doi.org/10.1145/2001858.2002141

T. Bartz-Beielstein, M. Zaefferer, Model-based methods for continuous and discrete global optimization. Appl. Soft Comput. **55**, 154–167 (2017). ISSN 1568-4946. https://doi.org/10.1016/j.asoc.2017.01.039. http://www.sciencedirect.com/science/article/pii/S1568494617300546

T. Bartz-Beielstein, K.E. Parsopoulos, M.N. Vrahatis, Analysis of particle swarm optimization using computational statistics, in *Proceedings International Conference Numerical Analysis and Applied Mathematics (ICNAAM)*, Weinheim, Germany, ed. by T.E. Simos, Tsitouras C.H. (Wiley-VCH, 2004), pp. 34–37

T. Bartz-Beielstein, C. Lasarczyk, M. Preuss, Sequential parameter optimization, in *Proceedings 2005 Congress on Evolutionary Computation (CEC'05), Edinburgh, Scotland*, Piscataway NJ, ed. by B McKay et al. (IEEE Press, 2005), pp. 773–780. ISBN 0-7803-9363-5. https://doi.org/10.1109/CEC.2005.1554761

T. Bartz-Beielstein, M. Chiarandini, L. Paquete, M. Preuss (eds.), *Experimental Methods for the Analysis of Optimization Algorithms* (Springer, Berlin, Heidelberg, New York, 2010)

T. Bartz-Beielstein, M. Friese, M. Zaefferer, B. Naujoks, O. Flasch, W. Konen, P. Koch, Noisy optimization with sequential parameter optimization and optimal computational budget allocation, in *Proceedings of the 13th annual conference companion on Genetic and evolutionary computation*, New York, NY, USA (ACM, 2011), pp. 119–120. ISBN 978-1-4503-0690-4. https://doi.org/10.1145/2001858.2001926

T. Bartz-Beielstein, J. Branke, J. Mehnen, O. Mersmann, Evolutionary algorithms. *Wiley Interdisciplinary Reviews: Data Mining and Knowledge Discovery*, vol. 4(3), pp. 178–195 (2014). ISSN 1942-4795. https://doi.org/10.1002/widm.1124

T. Bartz-Beielstein, L. Gentile, M. Zaefferer, *In a Nutshell: Sequential Parameter Optimization* (2017). https://arxiv.org/abs/1712.04076v1

T. Bartz-Beielstein, M. Zaefferer, Q.C. Pham, Optimization via multimodel simulation. Struct. Multidiscip. Opt. **58**(3), 919–933 (2018). ISSN 1615-1488. https://doi.org/10.1007/s00158-018-1934-2

T. Bartz-Beielstein, C. Doerr, J. Bossek, S. Chandrasekaran, T. Eftimov, A. Fischbach, P. Kerschke, M. Lopez-Ibanez, K.M. Malan, J.H. Moore, B. Naujoks, P. Orzechowski, V. Volz, M. Wagner, T. Weise, *Benchmarking in Optimization: Best Practice and Open Issues*, Jul. 2020a. arxiv:2007.03488

T. Bartz-Beielstein, M. Dröscher, A. Gür, A. Hinterleitner, T. Lawton, O. Mersmann, D. Peeva, L. Reese, N. Rehbach, F. Rehbach, A. Sen, A. Subbotin, M. Zaefferer, Optimization and adaptation of a resource planning tool for hospitals under special consideration of the covid-19 pandemic, in *2021 IEEE Congress on Evolutionary Computation (CEC)*, pp. 728–735 (2021b). https://doi.org/10.1109/CEC45853.2021.9504732

T. Bartz-Beielstein, J. Stork, M. Zaefferer, M. Rebolledo, C. Lasarczyk, F. Rehbach, Spot: sequential parameter optimization toolbox v2.3.0. online (2021c). https://cran.r-project.org/package=SPOT. Accessed 15 Mar. 2021

T. Bartz-Beielstein, M. Zaefferer, F. Rehbach, *In a Nutshell—The Sequential Parameter Optimization Toolbox*, Dec. 2021. arXiv:1712.04076

T. Bartz-Beielstein et al., *Benchmarking—Best Practice and Open Issues* (2020b)

M. Beck, F. Dumpert, J. Feuerhake, *Machine Learning in Official Statistics*, pp. 1–66 (2020). arxiv:1812.10422

Y. Bengio, Practical recommendations for gradient-based training of deep architectures, in *Neural Networks: Tricks of the Trade*, volume 7700 of *Lecture Notes in Computer Science*, ed. by G. Montavon, K.-R. Müller, G.B. Orr (Springer, 2012), pp. 437–478. https://doi.org/10.1007/978-3-642-35289-8_26

J. Bergstra, D. Yamins, D.D. Cox, Making a science of model search: hyperparameter optimization in hundreds of dimensions for vision architectures, in *Proceedings of the 30th International Conference on International Conference on Machine Learning—ICML'13*, vol. 28, pp. I–115–I–123 (2013). JMLR.org

J. Bergstra, Y. Bengio, Random search for hyper-parameter optimization. J. Mach. Learn. Res. **13**, 281–305 (2012). ISSN 1532-4435; 1533-7928/e

A. Biedenkapp, J. Marben, M. Lindauer, F. Hutter, Cave: configuration assessment, visualization and evaluation, in *Proceedings of the International Conference on Learning and Intelligent Optimization (LION'18)*, June 2018

M. Birattari, Z. Yuan, P. Balaprakash, T. Stützle, F-race and iterated F-race: an overview, in *Empirical Methods for the Analysis of Optimization Algorithms*. ed. by T. Bartz-Beielstein, M. Chiarandini, L. Paquete, M. Preuss (Springer, Berlin, Heidelberg, New York, 2009)

B. Bischl, M. Binder, M. Lang, T. Pielok, J. Richter, S. Coors, J. Thomas, T. Ullmann, M. Becker, A. Boulesteix, D. Deng, M. Lindauer. Hyperparameter optimization: foundations, algorithms, best practices and open challenges (2021a). arXiv:2107.05847 [stat.ML]

B. Bischl, M. Lang, L. Kotthoff, J. Schiffner, J. Richter, E. Studerus, G. Casalicchio, Z.M. Jones, MLR: machine learning in R. J. Mach. Learn. Res. **17**(170), 1–5 (2016). https://jmlr.org/papers/v17/15-066.html

B. Bischl, M. Binder, M. Lang, T. Pielok, J. Richter, S. Coors, J. Thomas, T. Ullmann, M. Becker, A.-L. Boulesteix, D. Deng, M Lindauer, Hyperparameter optimization: foundations, algorithms, best practices and open challenges, pp. 1–70 (2021b). arxiv:2107.05847

L. Bloch, C.M. Friedrich, Using bayesian optimization to effectively tune random forest and xgboost hyperparameters for early alzheimer's disease diagnosis, in *Wireless Mobile Communication and Healthcare*, Cham, ed. by J. Ye, M.J. O'Grady, G. Civitarese, K. Yordanova (Springer International Publishing, 2021), pp. 285–299. ISBN 978-3-030-70569-5

P. Bönisch, R. Inderst, Using the statistical concept of "severity" to assess seemingly contradictory statistical evidence (with a particular application to damage estimation). LawFin Working Paper 3, Center for Advanced Studies on the Foundations of Law and Finance (LawFin), Goethe University (2020)

A.J. Booker, J.E. Dennis Jr, P.D. Frank, D.B. Serafini, V. Torczon, M.W. Trosset, A rigorous framework for optimization of expensive functions by surrogates. Struct. Optim. **17**(1), 1–13 (1999) https://doi.org/10.1007/BF01197708. URL http://link.springer.com/10.1007/BF01197708

L. Breiman, Random forests. Mach. Learn. **45**(1), 5–32 (2001)

Brownlee, J., XGBoost with python—gradient boosted trees with XGBoost and scikit-learn. *Machine Learning Mastery*, v1.10 ed. (2018)

A. Bunte, A. Fischbach, J. Strohschein, T. Bartz-Beielstein, H. Faeskorn-Woyke, O. Niggemann, Evaluation of cognitive architectures for cyber-physical production systems, in *24th IEEE International Conference on Emerging Technologies and Factory Automation, ETFA 2019, Zaragoza, Spain*, 10–13 Sept. 2019, pp. 729–736 (2019). https://doi.org/10.1109/ETFA.2019.8869038

R.E. Caflisch, W. Morokoff, A. Owen, Valuation of mortgage backed securities using brownian bridges to reduce effective dimension (1997)

J. Chambers, W. Cleveland, B. Kleiner, P. Tukey, *Graphical Methods for Data Analysis* (Wadsworth, Belmont, CA, 1983)

J.M. Chambers, T.H. Hastie (eds.), *Statistical Models in S* (Wadsworth and Brooks/Cole, Pacific Grove, CA, 1992)

C.-C. Chang, C.-J. Lin, LIBSVM. ACM Trans. Intell. Syst. Technol. **2**(3), 1–27 (2011). https://doi.org/10.1145/1961189.1961199. Updated version from November 29, 2019

C.H. Chen, *Stochastic Simulation Optimization: An Optimal Computing Budget Allocation.* (World Scientific, 2010)

T. Chen, C. Guestrin, *XGBoost: A Scalable Tree Boosting System* (2016). arxiv:1603.02754

T. Chen, T. He, M. Benesty, V. Khotilovich, Y. Tang, H. Cho, K. Chen, R. Mitchell, I. Cano, T. Zhou, M. Li, J. Xie, M. Lin, Y. Geng, Y. Li, Package xgboost (reference manual, v1.2.0.1). online (2020). https://cran.r-project.org/web/packages/xgboost/xgboost.pdf

V. Cherkassky, Y. Ma, Practical selection of SVM parameters and noise estimation for SVM regression. Neural Netw. **17**(1), 113–126 (2004). https://doi.org/10.1016/s0893-6080(03)00169-2

M. Chiarandini, Y. Goegebeur, *Mixed Models for the Analysis of Optimization Algorithms* (Springer, Berlin, Heidelberg, 2010), pp. 225–264. ISBN 978-3-642-02538-9. https://doi.org/10.1007/978-3-642-02538-9_10

D. Choi, C.J. Shallue, Z. Nado, J. Lee, C.J. Maddison, G.E. Dahl, *On Empirical Comparisons of Optimizers for Deep Learning* (2019). arxiv:1910.05446

F. Chollet, J.J. Allaire, *Deep Learning with R* (Manning, 2018)

J. Cohen, Copyright. Revised ed. (Statistical Power Analysis for the Behavioral Sciences, 1977)

P.R. Cohen *Empirical Methods for Artificial Intelligence* (MIT Press, Cambridge MA, 1995)

W.D. Cook, M. Kress, *Ordinal Information and Preference Structures: Decision Models and Applications* (Prentice-Hall, Inc., USA, 1992). ISBN 0136301207

T. Cover, P. Hart, Nearest neighbor pattern classification. IEEE Trans. Inf. Theory **13**(1), 21–27 (1967). https://doi.org/10.1109/tit.1967.1053964

J. Cuesta Ramirez, R. Le Riche, O. Roustant, G. Perrin, C. Durantin, A. Glière, A comparison of mixed-variables bayesian optimization approaches. Adv. Model. Simul. Eng. Sci **9**(1) (2022). https://doi.org/10.1186/s40323-022-00218-8

G. De Ath, R.M. Everson, A.A.M. Rahat, J.E. Fieldsend, Greed is good: exploration and exploitation trade-offs in bayesian optimisation, Nov. 2019

J.-C. de Borda, Mémoire sur les élections au scrutin. *Histoire de l'Académie Royale des Sciences* (1781)

Q.H. Doan, D.-K. Thai, N.L. Tran, A hybrid model for predicting missile impact damages based on k-nearest neighbors and Bayesian optimization. J. Sci. Technol. Civil Eng. (STCE)-NUCE **14**(3), 1–14 (2020). https://doi.org/10.31814/stce.nuce2020-14(3)-01

C. Doerr, H. Wang, F. Ye, S. van Rijn, T. Bäck, IOHprofiler: A Benchmarking and Profiling Tool for Iterative Optimization Heuristics, Oct 2018. arXiv:1810.05281

P. Domingos, A few useful things to know about machine learning. Commun. ACM **55**(10), 78–87 (2012). https://doi.org/10.1145/2347736.2347755. (October)

X. Dong, M. Tan, A.W. Yu, D. Peng, B. Gabrys, Q.V. Le, *AutoHAS: Efficient Hyperparameter and Architecture Search* (2020). arxiv:2006.03656

T. Dozat, *Incorporating Nesterov Momentum into Adam* (2016)

J. Drozdal, J. Weisz, D. Wang, G. Dass, B. Yao, C. Zhao, M. Muller, L. Ju, H. Su, Trust in automl: exploring information needs for establishing trust in automated machine learning systems, in *Proceedings of the 25th International Conference on Intelligent User Interfaces*, IUI '20, New York, NY, USA (Association for Computing Machinery, 2020), pp. 297–307. ISBN 9781450371186. https://doi.org/10.1145/3377325.3377501

H. Drucker, C. Cortes, Boosting decision trees, in *NIPS 95: Proceedings of the 8th International Conference on Neural Information Processing Systems*, pp. 479–485, Nov. 1995

J. Duchi, E. Hazan, Y. Singer, Adaptive subgradient methods for online learning and stochastic optimization. J. Mach. Learn. Res. **12**(61), 2121–2159 (2011). http://jmlr.org/papers/v12/duchi11a.html

F. Dumpert, M. Beck, Einsatz von Machine-Learning-Verfahren in amtlichen Unternehmensstatistiken. AStA Wirtschafts- und Sozialstatistisches Archiv **11**, 83–106 (2017)

F. Dumpert, M. Beck, Verbesserung der Datengrundlage der Mindestlohnforschung mittels maschineller Lernverfahren. Submitted (2021)

F. Dumpert, K. von Eschwege, M. Beck, Einsatz von Support Vector Machines bei der Sektorzuordnung von Unternehmen. WISTA - Wirtschaft und Statistik 1(2016), 87–97 (2016)

B. Efron, R.J. Tibshirani, *An Introduction to the Bootstrap* (Chapman and Hall, London, 1993)

K. Eggensperger, M. Feurer, F. Hutter, J. Bergstra, J. Snoek, H.H. Hoos, K. Leyton-Brown, Towards an empirical foundation for assessing bayesian optimization of hyperparameters, in *In NIPS Workshop on Bayesian Optimization in Theory and Practice* (2013)

T. Elsken, J. Metzen, F. Hutter, Neural Archit. Sear.: A Surv. 20(55), 1–21 (2019)

M.T.M. Emmerich, *Single- and Multi-objective Evolutionary Design Optimization: Assisted by Gaussian Random Field Metamodels.* Ph.D. thesis, Universität Dortmund, Germany (2005). http://hdl.handle.net/2003/21807

N. Erickson, J. Mueller, A. Shirkov, H. Zhang, P. Larroy, M. Li, A. Smola, *AutoGluon-Tabular: Robust and Accurate AutoML for Structured Data* (2020). arxiv:2003.06505

S. Falkner, A. Klein, F. Hutter, *BOHB: Robust and Efficient Hyperparameter Optimization at Scale* (2018). arxiv:1807.01774

J. Feuerhake, F. Dumpert, Erkennung nicht relevanter Unternehmen in den Handwerksstatistiken. WISTA - Wirtschaft und Statistik 2(2016), 79–94 (2016)

J. Feuerhake, K. Lange, A. Siegismund, E. Vigneau, Kodierung des Geburtsstaats in der Wanderungsstatistik. WISTA - Wirtschaft und Statistik 3(2020), 98–110 (2020)

M. Feurer, K. Eggensperger, S. Falkner, M. Lindauer, F. Hutter. Auto-Sklearn 2.0: Hands-free AutoML via Meta-Learning, July 2020. arXiv:2007.04074

R.A. Fisher, *Statistical Methods for Research Workers* (Edinburgh Oliver & Boyd, 1925)

R. Fletcher, C.M. Reeves, Function minimization by conjugate gradients. Comput. J. 7(2), 149–154 (1964). ISSN 0010-4620. https://doi.org/10.1093/comjnl/7.2.149

A. Forrester, A. Sóbester, A. Keane, *Engineering Design via Surrogate Modelling* (Wiley, 2008a)

A. Forrester, A. Sobester, A. Keane, *Engineering Design via Surrogate Modelling* (Wiley, 2008b)

F. Freudenstein, B. Roth, Numerical solution of systems of nonlinear equations. J. ACM 10(4), 550–556 (1963). ISSN 0004-5411. https://doi.org/10.1145/321186.321200

Y. Freund, R.E. Schapire, A decision-theoretic generalization of on-line learning and an application to boosting. J. Comput. Syst. Sci. 55(1), 119–139 (1997). https://doi.org/10.1006/jcss.1997.1504

J. Friedman, T. Hastie, R. Tibshirani, Regularization paths for generalized linear models via coordinate descent. J. Stat. Softw. 33(1) (2010). https://doi.org/10.18637/jss.v033.i01

J. Friedman, T. Hastie, R. Tibshirani, B. Narasimhan, K. Tay, N. Simon, J. Qian, Package glmnet (reference manual, v4.0-2). https://cran.r-project.org/web/packages/glmnet/glmnet.pdf. Accessed 22 Nov. 2020

J.H. Friedman, Greedy function approximation: a gradient boosting machine. Ann. Stat. 29(5), 1189–1232 (2001). ISSN 00905364. http://www.jstor.org/stable/2699986

J.H. Friedman, Stochastic gradient boosting. Comput. Stat. Data Anal. 38(4), 367–378 (2002). https://doi.org/10.1016/s0167-9473(01)00065-2. (Feb)

J.H. Friedman, J.L. Bentley, R.A. Finkel, An algorithm for finding best matches in logarithmic expected time. ACM Trans. Math. Softw. 3(3), 209–226 (1977). ISSN 0098-3500. https://doi.org/10.1145/355744.355745

L. Gentile, T. Bartz-Beielstein, M. Zaefferer, *Sequential Parameter Optimization for Mixed-Discrete Problems* (Springer International Publishing, Cham, 2021), pp. 333–355. ISBN 978-3-030-60166-9. https://doi.org/10.1007/978-3-030-60166-9_10

J.E. Gentle, W.H.ärdle, Y. Mori, Computational statistics: an introduction, in *Computational Statistics*, ed. by J.E. Gentle, W. Härdle, Y. Mori (Springer, Berlin, Heidelberg, New York, 2004), pp. 3–16

P. Gijsbers, F. Pfisterer, J.N. van Rijn, B. Bischl, J. Vanschoren, *Meta-learning for Symbolic Hyperparameter Defaults*, pp. 1–13 (2021). https://arxiv.org/abs/2106.05767

C. Gillespie, Package benchmarkme (reference manual, v1.0.5). https://cran.r-project.org/web/packages/ranger/ranger.pdf. Accessed 09 Feb. 2021

Y. Goldberg, A primer on neural network models for natural language processing. J. Artif. Intell. Res **57**, 345–420 (2016). https://doi.org/10.1613/jair.4992. (November)

R. Gomes Mantovani, T. Horváth, R. Cerri, S. Barbon Junior, J. Vanschoren, A. Carlos P. de Leon Ferreira de Carvalho, An empirical study on hyperparameter tuning of decision trees (2018). arXiv:1812.02207, v2

R.B. Gramacy, *Surrogates* (CRC Press, 2020)

N. Guenther, M. Schonlau, Support vector machines. Stata J.: Promot. Commun. Stat. Stata **16**(4), 917–937 (2016). https://doi.org/10.1177/1536867x1601600407. (December)

R.T. Haftka, Requirements for papers focusing on new or improved global optimization algorithms. Struct. Multidiscip. Optim. **54**(1), 1–1 (2016). ISSN 1615-1488. https://doi.org/10.1007/s00158-016-1491-5

N. Hansen, The CMA evolution strategy: a comparing review, in *Towards a New Evolutionary Computation. Advances on Estimation of Distribution Algorithms*, ed. by J.A. Lozano, P. Larranaga, I. Inza, E. Bengoetxea (Springer, 2006), pp. 75–102

A. Hart, Mann-whitney test is not just a test of medians: differences in spread can be important. Bmj **323**(7309), 391–393 (2001)

N. Hasebrook, F. Morsbach, N. Kannengießer, J. Franke, F. Hutter, A. Sunyaev, Why Do Machine Learning Practitioners Still Use Manual Tuning? A Qualitative Study, Mar. 2022. arXiv:2203.01717

T. Hastie. *The Elements of Statistical Learning: Data Mining, Inference, and Prediction* (Springer, New York, 2nd ed., 2009). ISBN 9780387848570

T. Hastie, J. Qian, An introduction to glmnet (2016). https://cran.r-project.org/web/packages/glmnet/vignettes/glmnet.pdf. Accessed 22 Nov. 2020

T. Hastie, R. Tibshirani, J. Friedman. *The Elements of Statistical Learning* (Springer, 2nd ed., 2017). https://doi.org/10.1007/978-0-387-84858-7. 12th printing

K. Hechenbichler, K. Schliep, Weighted k-nearest-neighbor techniques and ordinal classification. Discussion Paper 399, Ludwig-Maximilians University Munic (2004)

L. Hedges, I. Olkin, *Statistical Methods for Meta-Analysis* (Academic Press, Orlando, FL, 1985)

K. Hornik, D. Meyer, Deriving consensus rankings from benchmarking experiments, in *Advances in Data Analysis*, ed. by R. Decker, H.J. Lenz (Springer, Berlin, Heidelberg, 2007), pp. 163–170. ISBN 978-3-540-70981-7

C.-W. Hsu, C.-C. Chang, C.-J. Lin, A practical guide to support vector classication (2016). https://www.csie.ntu.edu.tw/~cjlin/papers/guide/guide.pdf. Accessed 29 Nov. 2020

F. Hutter, H. H. Hoos, K. Leyton-Brown, Sequential model-based optimization for general algorithm configuration (extended version). Technical Report TR-2010-10, University of British Columbia, Department of Computer Science (2010a). http://www.cs.ubc.ca/~hutter/papers/10-TR-SMAC.pdf

F. Hutter, L. Kotthoff, J. Vanschoren, eds., *Automated Machine Learning: Methods, Systems, Challenges* (Springer, 2019). http://automl.org/book

F. Hutter, T. Bartz-Beielstein, H. Hoos, K. Leyton-Brown, K.P. Murphy, Sequential model-based parameter optimisation: an experimental investigation of automated and interactive approaches, in *Experimental Methods for the Analysis of Optimization Algorithms*, ed. by T. Bartz-Beielstein, M. Chiarandini, L. Paquete, M. Preuss (Springer, Berlin, Heidelberg, New York, 2010b), pp. 361–414

G. James, D. Witten, T. Hastie, R. Tibshirani, *An Introduction to Statistical Learning with Applications in R* (Springer, 7th ed., 2014)

G. James, D. Witten, T. Hastie, R. Tibshirani, *An Introduction to Statistical Learning* (Springer, 7th ed., 2017). https://doi.org/10.1007/978-1-4614-7138-7

K. Janocha, W.M. Czarnecki, On loss functions for deep neural networks in classification. Schedae Informaticae, 1/2016 (2017). https://doi.org/10.4467/20838476si.16.004.6185

H. Jin, Q. Song, X. Hu, Auto-keras: an efficient neural architecture search system, in *Proceedings of the 25th ACM SIGKDD International Conference on Knowledge Discovery and Data Mining*, KDD'19, New York, NY, USA (Association for Computing Machinery, 2019), pp. 1946–1956.

ISBN 9781450362016. https://doi.org/10.1145/3292500.3330648. URL https://doi.org/10.1145/3292500.3330648

D.R. Jones, M. Schonlau, W.J. Welch, Efficient global optimization of expensive black-box functions. J. Glob. Optim. **13**, 455–492 (1998)

I. Karmanov, M. Salvaris, M. Fierro, D. Dean, Comparing deep learning frameworks: a rosetta stone approach. *Microsoft docs, Microsoft* (2018). https://docs.microsoft.com/de-de/archive/blogs/machinelearning/comparing-deep-learning-frameworks-a-rosetta-stone-approach. Accessed 29 May 2021

D. Jacob Kedziora, K. Musial, B. Gabrys, *AutonoML: Towards an Integrated Framework for Autonomous Machine Learning* (2020). http://arxiv.org/abs/2012.12600

J.G. Kemeny, Mathematics without numbers. Daedalus **88**(4), 577–591 (1959). http://www.jstor.org/stable/20026529

J.G. Kemeny, J.L. Snell, *Preference Ranking: An Axiomatic Approach* (MIT Press, Cambridge, MA, 1962)

M.G. Kendall, A new measure of rank correlation. Biometrika **30**(1–2), 81–93 (1938)

F. Khan, S. Kanwal, S. Alamri, B. Mumtaz, Hyper-parameter optimization of classifiers, using an artificial immune network and its application to software bug prediction. IEEE Access **8**, 20954–20964 (2020). https://doi.org/10.1109/access.2020.2968362

D.P. Kingma, J. Ba, *Adam: A Method for Stochastic Optimization*, Dec. 2014. arXiv:1412.6980

D.P. Kingma, J. Ba, Adam: a method for stochastic optimization, in *3rd International Conference on Learning Representations, ICLR 2015, San Diego, CA, USA, May 7-9, 2015, Conference Track Proceedings*, ed. by Y. Bengio, Y. LeCun (2015). http://arxiv.org/abs/1412.6980

Kleijnen, Experimental design for sensitivity analysis, optimization, and validation of simulation models, in *Handbook of Simulation*, ed. by J. Banks (Wiley, New York NY, 1997)

J.P.C. Kleijnen, *Statistical Tools for Simulation Practitioners* (Marcel Dekker, New York, NY, 1987)

W. Koehrsen, An introductory example of bayesian optimization in python with hyperopt (2018). https://towardsdatascience.com/an-introductory-example-of-bayesian-optimization-in-python-with-hyperopt-aae40fff4ff0

R. Kohavi, A study of cross-validation and bootstrap for accuracy estimation and model selection, in *Proceedings of the 14th International Joint Conference on Artificial Intelligence—IJCAI'95*, San Francisco, CA, USA, vol. 2 (Morgan Kaufmann Publishers Inc., 1995), pp. 1137–1143. ISBN 1558603638

L. Kotthoff, C. Thornton, H.H. Hoos, F. Hutter, K. Leyton-Brown, Auto-weka 2.0: automatic model selection and hyperparameter optimization in weka. J. Mach. Learn. Res. **18**(25), 1–5 (2017). http://jmlr.org/papers/v18/16-261.html

Alex Krizhevsky, Ilya Sutskever, and Geoffrey E Hinton. Imagenet classification with deep convolutional neural networks. In *Advances in Neural Information Processing Systems*, vol. 25, ed. by F. Pereira, C.J. Burges, L. Bottou, K.Q. Weinberger (Curran Associates, Inc., 2012). https://proceedings.neurips.cc/paper/2012/file/c399862d3b9d6b76c8436e924a68c45b-Paper.pdf

W.H. Kruskal, W Allen Wallis, Use of ranks in one-criterion variance analysis. J. Am. Stat. Assoc. **47**(260), 583–621 (1952)

Upmanu Lall and Ashish Sharma. A nearest neighbor bootstrap for resampling hydrologic time series. Water Res. Res. **32**(3), 679–693 (1996). https://doi.org/10.1029/95wr02966

S. Leary, A. Bhaskar, A. Keane, Optimal orthogonal-array-based latin hypercubes. J. Appl. Stat. **30**(5), 585–598 (2003). https://doi.org/10.1080/0266476032000053691

Y.A. LeCun, L. Bottou, G.B. Orr, K.-R. Müller, Efficient BackProp, in *Neural Networks: Tricks of the Trade*, volume 7700 of *Lecture Notes in Computer Science* (Springer, 2012), pp. 9–48. https://doi.org/10.1007/978-3-642-35289-8_3

R.M. Lewis, V. Torczon, M.W. Trosset, Direct search methods: then and now. J. Comput. Appl. Math. **124**(1–2), 191–207 (2000)

L. Li, A. Talwalkar, Random search and reproducibility for neural architecture search. *Conference on Uncertainty in Artificial Intelligence (UAI)* (2019). http://arxiv.org/abs/1902.07638

L. Li, K. Jamieson, G. DeSalvo, A. Rostamizadeh, A. Talwalkar, *Hyperband: A Novel Bandit-Based Approach to Hyperparameter Optimization*, Mar. 2016. arXiv:1603.06560

X. Li, T. Long, G. Gary Wang, K. Haji Hajikolaei, R. Shi, [rbf]-based high dimensional model representation method using proportional sampling strategy, in *Advances in Structural and Multidisciplinary Optimization*, Cham, ed. by A. Schumacher, T. Vietor, S. Fiebig, K.-U. Bletzinger, K. Maute (Springer International Publishing, 2018), pp. 259–268. ISBN 978-3-319-67988-4

R. Liaw, E. Liang, R. Nishihara, P. Moritz, J.E. Gonzalez, I. Stoica, Tune: a research platform for distributed model selection and training (2018). arXiv:1807.05118

T.P. Lillicrap, J.J. Hunt, A. Pritzel, N. Heess, T. Erez, Y. Tassa, D. Silver, D. Wierstra, Continuous control with deep reinforcement learning, Sept. 2015. arXiv:1509.02971

M. Lindauer, F. Hutter, Best practices for scientific research on neural architecture search. J. Mach. Learn. Res. **21**(243), 1–18 (2020)

M. Lindauer, K. Eggensperger, M. Feurer, A. Biedenkapp, D. Deng, C. Benjamins, T. Ruhkopf, R. Sass, F. Hutter, Smac3: a versatile bayesian optimization package for hyperparameter optimization. J. Mach. Learn. Res **23**(54), 1–9 (2022). http://jmlr.org/papers/v23/21-0888.html

H.R. Lindman, *Analysis of Variance in Complex Experimental Designs* (WH Freeman & Co, 1974)

B. Liu, *A Very Brief and Critical Discussion on AutoML* (2018). arxiv:1811.03822

F.G. Lobo, C.F. Lima, Z. Michalewicz eds. *Parameter Setting in Evolutionary Algorithms*, volume 54 of *Studies in Computational Intelligence* (Springer, 2007). ISBN 978-3-540-69431-1

M. López-Ibáñez, J. Dubois-Lacoste, L. Pérez Cáceres, M. Birattari, T. Stützle, Iterated racing for automatic algorithm configuration. The irace package. Oper. Res. Perspect. **3**, 43–58 (2016). https://doi.org/10.1016/j.orp.2016.09.002

M. López-Ibáñez, J. Dubois-Lacoste, L. Pérez Cáceres, T. Stützle, M. Birattari, The irace package: iterated racing for automatic algorithm configuration. Oper. Res. Perspect. **3**, 43–58 (2016). https://doi.org/10.1016/j.orp.2016.09.002

M. López-Ibáñez, J. Branke, L. Paquete, *Reproducibility in Evolutionary Computation* (2021a). arxiv:2102.03380

M. López-Ibáñez, J. Branke, L. Paquete, Reproducibility in evolutionary computation. ACM Trans. Evol. Learn. Optim. **1**(4) (2021b). ISSN 2688-299X. https://doi.org/10.1145/3466624

A.C. Lorena, L.P.F. Garcia, J. Lehmann, M.C.P. Souto, T.K. Ho, How complex is your classification problem? A survey on measuring classification complexity **52**(5) (2019). ISSN 0360-0300. https://doi.org/10.1145/3347711

G. Louppe, Understanding random forests—from theory to practice (2015)

Y. Mack, T. Goel, W. Shyy, R. Haftka, *Surrogate Model-Based Optimization Framework: A Case Study in Aerospace Design*. (Springer, Berlin, Heidelberg, 2007), pp. 323–342. ISBN 978-3-540-49774-5. https://doi.org/10.1007/978-3-540-49774-5_14

R.G. Mantovani, A.L.D. Rossi, J. Vanschoren, B. Bischl, A.C.P.L.F. de Carvalho, Effectiveness of random search in SVM hyper-parameter tuning, in *2015 International Joint Conference on Neural Networks (IJCNN)* (IEEE, 2015). https://doi.org/10.1109/ijcnn.2015.7280664

D.G. Mayo, *Error and the Growth of Experimental Knowledge* (The University of Chicago Press, Chicago IL, 1996)

D.G. Mayo, *Statistical Inference as Severe Testing: How to Get Beyond the Statistics Wars* (Cambridge University Press, 2018). https://doi.org/10.1017/9781107286184

D.G. Mayo, A. Spanos, Severe testing as a basic concept in a neyman-pearson philosophy of induction. Br. J. Philos. Sci. **57**(2), 323–357 (2006). https://doi.org/10.1093/bjps/axl003. http://bjps.oxfordjournals.org/cgi/content/abstract/57/2/323

H. Mazzawi X. Gonzalvo, Introducing model search: an open source platform for finding optimal ML models, Feb 2021. https://ai.googleblog.com/2021/02/introducing-model-search-open-source.html?m=1

H. Mazzawi, X. Gonzalvo, A. Kracun, P. Sridhar, N.A. Subrahmanya, I. Lopez-Moreno, H. Park, P. Violette, Improving keyword spotting and language identification via neural architecture search at scale, in *INTERSPEECH* (2019)

W.S. McCulloch, W. Pitts, A logical calculus of the ideas immanent in nervous activity. Bull. Math. Biophys. **5**(4), 115–133 (1943). https://doi.org/10.1007/BF02478259

D. Meignan, S. Knust, J.-M. Frayet, G. Pesant, N. Gaud, A review and taxonomy of interactive optimization methods in operations research. ACM Trans. Int. Intell. Syst. (2015)

H. Mendoza, A. Klein, M. Feurer, J. Tobias Springenberg, F. Hutter, Towards automatically-tuned neural networks, in *ICML 2016 AutoML Workshop*, volume 64 of *JMLR: Workshop and Conference Proceedings*, pp. 58–65 (2016)

H. Mendoza, A. Klein, M. Feurer, J.T. Springenberg, M. Urban, M. Burkart, M. Dippel, M. Lindauer, F. Hutter, *Towards Automatically-Tuned Deep Neural Networks* (Springer International Publishing, Cham, 2019), pp. 135–149. ISBN 978-3-030-05318-5. https://doi.org/10.1007/978-3-030-05318-5_7

G. Menghani, *Efficient Deep Learning: A Survey on Making Deep Learning Models Smaller, Faster, and Better*, June 2021

O. Mersmann, M. Preuss, H. Trautmann, Benchmarking evolutionary algorithms: towards exploratory landscape analysis, in *Parallel Problem Solving from Nature, PPSN XI*, Berlin, Heidelberg. ed. by R. Schaefer, C. Cotta, J. Kołodziej, G. Rudolph (Springer Berlin Heidelberg, 2010a), pp. 73–82. ISBN 978-3-642-15844-5

O. Mersmann, H. Trautmann, B. Naujoks, C. Weihs, Benchmarking evolutionary multiobjective optimization algorithms, in *IEEE Congress on Evolutionary Computation*, pp. 1–8 (2010b). https://doi.org/10.1109/CEC.2010.5586241

O. Mersmann, B. Bischl, H. Trautmann, M. Preuss, C. Weihs, G. Rudolph, Exploratory landscape analysis, in *Proceedings of the 13th Annual Conference on Genetic and Evolutionary Computation*, GECCO '11, New York, NY, USA (Association for Computing Machinery, 2011), pp. 829–836. ISBN 9781450305570. https://doi.org/10.1145/2001576.2001690

O. Mersmann, M. Preuss, H. Trautmann, B. Bischl, C. Weihs, Analyzing the BBOB results by means of benchmarking concepts. Evol. Comput. **23**(1), 161–185 (2015). ISSN 1063-6560. https://doi.org/10.1162/EVCO_a_00134

C.E. Metz, Basic priciples of ROC analysis. Sem. Nucl. Med. VIII **4**, 283–298 (1978). (October)

D. Meyer, K. Hornik, *Relations: Data Structures and Algorithms for Relations* (2022). https://cran.r-project.org/package=relations. R package version 0.6-12

D. Meyer, E. Dimitriadou, K. Hornik, A. Weingessel, F. Leisch, Package e1071 (reference manual, v1.7-4) (2020). https://cran.r-project.org/web/packages/e1071/e1071.pdf. Accessed 29 Nov. 2020

D.C. Montgomery, *Design and Analysis of Experiments*, 9th edn. (Wiley, New York, NY, 2017)

J. Moosbauer, J. Herbinger, G. Casalicchio, M. Lindauer, B. Bischl, *Explaining Hyperparameter Optimization Via Partial Dependence Plots*, pp. 1–21 (2021). arxiv:2111.04820

J.J. More, B.S. Garbow, K.E. Hillstrom, Testing unconstrained optimization software. ACM Trans. Math. Softw. **7**(1), 17–41 (1981)

R.H. Myers, D.C. Montgomery, C.M. Anderson-Cook, *Response Surface Methodology: Process and Product Optimization using Designed Experiments* (Wiley, 2016)

J.A. Nelder, R. Mead, A simplex method for function minimization. Comput. J. **7**(4), 308–313 (1965). ISSN 0010-4620. https://doi.org/10.1093/comjnl/7.4.308

J. Neyman, *First Course in Probability and Statistics* (Henry Holt, New York, NY, 1950)

T. O'Malley, E. Bursztein, J. Long, F. Chollet, H. Jin, L. Invernizzi, et al., *Keras Tuner* (2019). https://github.com/keras-team/keras-tuner

H. Osman, M. Ghafari, O. Nierstrasz, Hyperparameter optimization to improve bug prediction accuracy, in *IEEE Workshop on Machine Learning Techniques for Software Quality Evaluation (MaLTeSQuE)*, vol. 02 (IEEE, 2017). https://doi.org/10.1109/maltesque.2017.7882014

R.R. Picard, R. Dennis Cook, Cross-validation of regression models. J. Am. Stat. Assoc. **79**(387), 575–583 (1984). https://doi.org/10.1080/01621459.1984.10478083. https://www.tandfonline.com/doi/abs/10.1080/01621459.1984.10478083

L. Prechelt, Early stopping—but when?, in *Neural Networks: Tricks of the Trade*, volume 7700 of *Lecture Notes in Computer Science*, G. Montavon, K.-R. Müller, G.B. Orr (Springer, 2012), pp. 53–67. https://doi.org/10.1007/978-3-642-35289-8_5

P. Probst, A.-L. Boulesteix, To tune or not to tune the number of trees in random forest. J. Mach. Learn. Res. **18**(181), 1–18 (2018). http://jmlr.org/papers/v18/17-269.html

P. Probst, B. Bischl, A.-L. Boulesteix, *Tunability: Importance of Hyperparameters of Machine Learning Algorithms* (2018). http://arxiv.org/abs/1802.09596

P. Probst, A.-L. Boulesteix, B. Bischl, Tunability: importance of hyperparameters of machine learning algorithms. J. Mach. Learn. Res. **20**(53), 1–32 (2019)

P. Probst, A.-L. Boulesteix, B. Bischl, Tunability: importance of hyperparameters of machine learning algorithms. J. Mach. Learn. Res. **20**(53), 1–32 (2019b). http://jmlr.org/papers/v20/18-444.html

P. Probst, M.N. Wright, A.-L. Boulesteix, Hyperparameters and tuning strategies for random forest. Wiley Interdiscip. Rev.: Data Min. Knowl. Discov. **9**(3), 01 (2019). https://doi.org/10.1002/widm.1301

R Core Team. *R: A Language and Environment for Statistical Computing* (R Foundation for Statistical Computing, Vienna, Austria, 2022). https://www.R-project.org

W.J. Radermacher, *Official Statistics 4.0* (Springer, 2020)

F. Rehbach, M. Zaefferer, B. Naujoks, T. Bartz-Beielstein, Expected improvement versus predicted value in surrogate-based optimization, in *Proceedings of the 2020 Genetic and Evolutionary Computation Conference*, GECCO'20, New York, NY, USA (Association for Computing Machinery, 2020), pp. 868–876. ISBN 9781450371285. https://doi.org/10.1145/3377930.3389816. URL https://doi.org/10.1145/3377930.3389816

E. Ridge, D. Kudenko, *Tuning an Algorithm Using Design of Experiments* (Springer, Berlin, Heidelberg, 2010), pp. 265–286. ISBN 978-3-642-02538-9. https://doi.org/10.1007/978-3-642-02538-9_11

D.A. Roberts, S. Yaida, B. Hanin, *The Principles of Deep Learning Theory, 2021*. arxiv:2106.10165

H.H. Rosenbrock, An automatic method for finding the greatest or least value of a function. Comput. J. **3**, 175–184 (1960)

S. Ruder, An overview of gradient descent optimization algorithms, Sept. 2017. arXiv:1609.04747

D.G. Saari, V.R. Merlin, A geometric examination of kemeny's rule. Soc. Choice Welf. **17**(3), 403–438 (2000). http://www.jstor.org/stable/41106368

T.J. Santner, B.J. Williams, W.I. Notz, *The Design and Analysis of Computer Experiments* (Springer, Berlin, Heidelberg, New York, 2003)

K. Schliep, K. Hechenbichler, A. Lizee, Package kknn (reference manual, v1.3.1) (2016). https://cran.r-project.org/web/packages/kknn/kknn.pdf. Accessed 29 Nov. 2020

E. Schmidt, Korrektur des Tätigkeitsschlüssels der Bundesagentur für Arbeit mithilfe maschineller Lernverfahren. WISTA - Wirtschaft und Statistik **6**(2020), 37–47 (2020)

R.M. Schmidt, F. Schneider, P. Hennig, *Descending through a Crowded Valley—Benchmarking Deep Learning Optimizers* (2020). arxiv:2007.01547

F. Schneider, L. Balles, P. Hennig, *DeepOBS: A Deep Learning Optimizer Benchmark Suite* (2019). arxiv:1903.05499

B. Schölkopf, A.J. Smola, *Learning with Kernels: Support Vector Machines, Regularization, Optimization, and Beyond* (MIT Press, Cambridge, MA, USA, 2001)

M Schonlau, *Computer Experiments and Global Optimization*. Ph.D. thesis, University of Waterloo, Ontario, Canada (1997)

P. Schratz, J. Muenchow, E. Iturritxa, J. Richter, A. Brenning, Hyperparameter tuning and performance assessment of statistical and machine-learning algorithms using spatial data. Ecol. Model. **406**, 109–120 (2019). https://doi.org/10.1016/j.ecolmodel.2019.06.002. (Aug)

E. Scornet, Tuning parameters in random forests. ESAIM: Proc. Surv. **60**, 144–162 (2017). https://doi.org/10.1051/proc/201760144

S. Senn, A comment on replication, p-values and evidence, s.n. goodman, statistics in medicine **11**, 875–879 (1992); Stat. Med. **21**(16), 2437–2444, Aug. 2002. ISSN 0277-6715 (Print); 0277-6715 (Linking). https://doi.org/10.1002/sim.1072

S. Senn. *Statistical Issues in Drug Development* (Wiley Blackwell, 3 ed., 2021)

C.J. Shallue, J. Lee, J. Antognini, J. Sohl-Dickstein, R. Frostig, G.E. Dahl, Measuring the effects of data parallelism on neural network training. J. Mach. Learn. Res. **20**, 1–49 (2019). arxiv:1811.03600

D.J. Sheskin, *Handbook of Parametric and Nonparametric Statistical Procedures* (Chapman and Hall/CRC, 2003)

F. Sigrist, *Gradient and Newton Boosting for Classication and Regression* (2020). arxiv:1808.03064, v7

N. Simon, J. Friedman, T. Hastie, R. Tibshirani, Regularization paths for coxs proportional hazards model via coordinate descent. J. Stat. Softw. **39**(5) (2011). https://doi.org/10.18637/jss.v039.i05

H. Singh, *Keras Tutorial: Deep Deterministic Policy Gradient (DDPG)* (2020). https://keras.io/examples/rl/ddpg_pendulum/. Accessed 18 Aug. 2021

J. Snoek, H. Larochelle, R.P. Adams, *Practical Bayesian Optimization of Machine Learning Algorithms*, June 2012. arXiv:1206.2944

A. Spanos, *Probability Theory and Statistical Inference* (Cambridge University Press, 1999)

N. Srivastava, G. Hinton, A. Krizhevsky, I. Sutskever, R. Salakhutdinov, Dropout: a simple way to prevent neural networks from overfitting. J. Mach. Learn. Res. **15**(56), 1929–1958 (2014). http://jmlr.org/papers/v15/srivastava14a.html

R. Storn, K. Price, Differential evolution—a simple and efficient heuristic for global optimization over continuous spaces. J. Glob. Optim. **11**(4), 341–359 (1997)

G. Strang, *Computational Science and Engineering* (Wellesley-Cambridge Press, USA, 2007)

C.H. Sudheer, R. Maheswaran, B.K. Panigrahi, S. Mathur, A hybrid SVM-PSO model for forecasting monthly streamflow. Neural Comput. Appl. **24**(6), 1381–1389 (2013). https://doi.org/10.1007/s00521-013-1341-y

I. Sutskever, J. Martens, G. Dahl, G. Hinton, On the importance of initialization and momentum in deep learning, in eds., *Proceedings of the 30th International Conference on Machine Learning*, volume 28 of *Proceedings of Machine Learning Research*, Atlanta, Georgia, USA, ed. by S. Dasgupta, D. McAllester, pp. 1139–1147, 17–19 June 2013. PMLR. https://proceedings.mlr.press/v28/sutskever13.html

V.A. Tatsis, K.E. Parsopoulos, Grid search for operator and parameter control in differential evolution, in *9th Hellenic Conference on Artificial Intelligence (SETN 2016)* (2016)

T.M. Therneau, E.J. Atkinson, An Introduction to Recursive Partitioning Using the RPART Routines. Technical report (2019)

T.M. Therneau, E.J. Atkinson, B. Ripley, Package rpart (reference manual, v4.1-15) (2019). https://cran.r-project.org/web/packages/rpart/rpart.pdf. Accessed 22 Nov. 2020

J. Thomas, S. Coors, B. Bischl, Automatic gradient boosting, in *International Workshop on Automatic Machine Learning at ICML* (2018)

C. Thornton, F. Hutter, H.H. Hoos, K. Leyton-Brown, Auto-weka: combined selection and hyperparameter optimization of classification algorithms, in *Proceedings of the 19th ACM SIGKDD International Conference on Knowledge Discovery and Data Mining*, KDD '13, New York, NY, USA (Association for Computing Machinery, 2013), pp. 847–855. ISBN 9781450321747. https://doi.org/10.1145/2487575.2487629

T. Tieleman, G. Hinton, Lecture 6.5-RMSProp: divide the gradient by running average of its recent magnitude **4**(2), 26–31 (2012)

J.W. Tukey, *Explorative data analysis* (Addison-Wesley, 1977). ISBN 978-0201076165. http://www.worldcat.org/title/exploratory-data-analysis/oclc/3058187

J.N. van Rijn, Frank Hutter. Hyperparameter importance across datasets, in *Proceedings of the 24th ACM SIGKDD International Conference on Knowledge Discovery & Data Mining* (ACM, July 2018). https://doi.org/10.1145/3219819.3220058

J. Vanschoren, Jan N. van Rijn, B. Bischl, L. Torgo, Openml: networked science in machine learning. SIGKDD Explor. **15**(2), 49–60 (2013). https://doi.org/10.1145/2641190.2641198

J. Vanschoren, J.N. van Rijn, B. Bischl, L. Torgo, OpenML: networked science in machine learning, July 2014. arXiv:1407.7722

A. Vodopija, J. Stork, T. Bartz-Beielstein, B. Filipič, Elevator group control as a constrained multiobjective optimization problem. Appl. Soft Comput. **115**, 108277 (2022). ISSN 1568-4946. https://doi.org/10.1016/j.asoc.2021.108277. https://www.sciencedirect.com/science/article/pii/S1568494621010899

C. Waibel, R. Evins, J. Carmeliet. Clustering and ranking based methods for selecting tuned search heuristic parameters, in *2019 IEEE Congress on Evolutionary Computation (CEC)*, pp. 2931–2940, June 2019. https://doi.org/10.1109/CEC.2019.8790261

H. Wang, D. Vermetten, F. Ye, C. Doerr, T. Bäck, Iohanalyzer: detailed performance analyses for iterative optimization heuristics. ACM Trans. Evol. Learn. Optim. **2**(1), 3:1–3:29 (2022). https://doi.org/10.1145/3510426

Y. Wang, Weiterentwicklung eines cfd-modells zur feuerraumsimulation, März (2019)

S. Wessing, M. Preuss, The true destination of EGO is multi-local optimization, in *2017 IEEE Latin American Conference on Computational Intelligence (LA-CCI)* (IEEE, Nov. 2017). https://doi.org/10.1109/la-cci.2017.8285677

A.C. Wilson, R. Roelofs, M. Stern, N. Srebro, B. Recht, The marginal value of adaptive gradient methods in machine learning, in *Advances in Neural Information Processing Systems*, vol. 30, ed. by I. Guyon, U. V. Luxburg, S. Bengio, H. Wallach, R. Fergus, S. Vishwanathan, R. Garnett (Curran Associates, Inc., 2017). https://proceedings.neurips.cc/paper/2017/file/81b3833e2504647f9d794f7d7b9bf341-Paper.pdf

M. Wistuba, A. Rawat, T. Pedapati, *A Survey on Neural Architecture Search* (2019). arxiv:1905.01392

D.H. Wolpert, W.G. Macready, No free lunch theorems for optimization. IEEE Trans. Evol. Comput. **1**(1), 67–82 (1997)

D.H. Wolpert, W.G. Macready, No free lunch theorems for optimization. IEEE Trans. Evol. Comput. **1**(1), 67–82 (1997). https://doi.org/10.1109/4235.585893

J. Wong, T. Manderson, M. Abrahamowicz, D.L. Buckeridge, R. Tamblyn, Can hyperparameter tuning improve the performance of a super learner? Epidemiology **30**(4), 521–53 (2019). https://doi.org/10.1097/ede.0000000000001027

M.N. Wright, Package ranger (reference manual, v0.12.1). https://cran.r-project.org/web/packages/ranger/ranger.pdf. Accessed 22 Nov. 2020

M.N. Wright, I.R. König, Splitting on categorical predictors in random forests. PeerJ **7**, e6339 (2019). https://doi.org/10.7717/peerj.6339

M.N. Wright, A. Ziegler, Ranger: a fast implementation of random forests for high dimensional data in c++ and r. J. Stat. Softw. Articl. **77**(1), 1–17 (2017). ISSN 1548-7660. https://doi.org/10.18637/jss.v077.i01

xgboost developers. xgboost release 1.2.1 (documentation) (2020). https://xgboost.readthedocs.io/_/downloads/en/stable/pdf/. Accessed 28- Nov. 2020

L. Yang, A. Shami, On hyperparameter optimization of machine learning algorithms: theory and practice. Neurocomputing **415**, 295–316 (2020). https://doi.org/10.1016/j.neucom.2020.07.061. (November)

K. Yu, C. Sciuto, M. Jaggi, C. Musat, M. Salzmann, Evaluating the search phase of neural architecture search, in *International Conference on Learning Representations* (2020). https://openreview.net/forum?id=H1loF2NFwr

M. Zaefferer, *Surrogate Models for Discrete Optimization Problems*. Ph.d. thesis, Technische Universität Dortmund, Dec. 2018. https://doi.org/10.17877/DE290R-19857

M. Zaefferer, CEGO: combinatorial efficient global optimization—v2.4.2. R Package, Online on CRAN (2021). https://cran.r-project.org/package=CEGO. Accessed 30 June 2022

M. Zaefferer, J. Stork, M. Friese, A. Fischbach, B. Naujoks, T. Bartz-Beielstein, Efficient global optimization for combinatorial problems, in *Proceedings of the Genetic and Evolutionary Com-*

putation Conference (GECCO'14), ed. by D.V. Arnold (ACM, 2014), pp. 871–878. https://doi.org/10.1145/2576768.2598282

M.D. Zeiler, ADADELTA: an adaptive learning rate method, Dec. 2012. arXiv:1212.5701

G. Zhang, L. Li, Z. Nado, J. Martens, S. Sachdeva, G.E. Dahl, C.J. Shallue, R. Grosse, *Which Algorithmic Choices Matter at Which Batch Sizes? Insights From a Noisy Quadratic Model* (2019). http://arxiv.org/abs/1907.04164

A. Zharmagambetov, S.S. Hada, M.A. Carreira-Perpinan, M. Gabidolla, An experimental comparison of old and new decision tree algorithms (2019). arXiv:1911.03054, v2

J. Zhou, Y. Qiu, S. Zhu, D. Jahed Armaghani, M. Khandelwal, E. Tonnizam Mohamad, Estimation of the TBM advance rate under hard rock conditions using XGBoost and bayesian optimization. *Underground Space*, July 2020. https://doi.org/10.1016/j.undsp.2020.05.008. Corrected proof

J. Zhou, J. Shi, G. Li, Fine tuning support vector machines for short-term wind speed forecasting. Energy Convers. Manag. **52**(4), 1990–1998 (2011). https://doi.org/10.1016/j.enconman.2010.11.007

L. Zimmer, M. Lindauer, F. Hutter, Auto-PyTorch tabular: multi-fidelity metalearning for efficient and robust AutoDL, June 2020. arXiv:2006.13799

B. Zoph, Q.V. Le, *Neural Architecture Search with Reinforcement Learning* (2016). arxiv:1611.01578

B. Zoph, V. Vasudevan, J. Shlens, Q.V. Le, *Learning Transferable Architectures for Scalable Image Recognition* (2017). arxiv:1707.07012

H. Zou, T. Hastie, Regularization and variable selection via the elastic net. J. R. Stat. Soc. Ser. B (Statistical Methodology) **67**(2), 301–320 (2005). ISSN 13697412, 14679868. http://www.jstor.org/stable/3647580

Index

© The Editor(s) (if applicable) and The Author(s) 2023
E. Bartz et al. (eds.), *Hyperparameter Tuning for Machine and Deep Learning with R*,
https://doi.org/10.1007/978-981-19-5170-1

Printed in the United States
by Baker & Taylor Publisher Services